ANNALS *of* THE NEW YORK ACADEMY OF SCIENCES

VOLUME
1280

ISBN-10: 1-57331-891-4; **ISBN-13:** 978-1-57331-891-4

ISSUE

Inositol Phospholipid Signaling in Physiology and Disease

T0344846

ISSUE EDITORS

William G. Kerr and Sandra Fernandes

State University of New York, Upstate Medical University, Syracuse, New York

This volume presents manuscripts stemming from the conference "Inositol Phospholipid Signaling in Physiology and Disease", which was supported by The New York Academy friend sponsors, Amgen, Aquinox Pharmaceuticals, Avanti Polar Lipids, Inc., Echelon Biosciences, and Oncothyreon, and by educational grants from Bristol-Myers Squibb, Gilead Sciences, and Infinity Pharmaceuticals, Inc.

TABLE OF CONTENTS

Annals of the New York Academy of Sciences (ISSN: 0077-8923 [print]; ISSN: 1749-6632 [online]) is published 30 times a year on behalf of the New York Academy of Sciences by Wiley Subscription Services, Inc., a Wiley Company, 111 River Street, Hoboken, NJ 07030-5774.

Mailing: *Annals of the New York Academy of Sciences* is mailed standard rate.

Postmaster: Send all address changes to ANNALS OF THE NEW YORK ACADEMY OF SCIENCES, Journal Customer Services, John Wiley & Sons Inc., 350 Main Street, Malden, MA 02148-5020.

Disclaimer: The publisher, the New York Academy of Sciences, and the editors cannot be held responsible for errors or any consequences arising from the use of information contained in this publication; the views and opinions expressed do not necessarily reflect those of the publisher, the New York Academy of Sciences, and editors, neither does the publication of advertisements constitute any endorsement by the publisher, the New York Academy of Sciences and editors of the products advertised.

Publisher: *Annals of the New York Academy of Sciences* is published by Wiley Periodicals, Inc., Commerce Place, 350 Main Street, Malden, MA 02148; Telephone: 781 388 8200; Fax: 781 388 8210.

Journal Customer Services: For ordering information, claims, and any inquiry concerning your subscription, please go to www.wileycustomerhelp.com/ask or contact your nearest office. *Americas:* Email: cs-journals@wiley.com; Tel:+1 781 388 8598 or 1 800 835 6770 (Toll free in the USA & Canada). *Europe, Middle East, Asia:* Email: cs-journals@wiley. com; Tel: +44 (0) 1865 778315. *Asia Pacific:* Email: cs-journals@wiley.com; Tel: +65 6511 8000. *Japan:* For Japanese speaking support, Email: cs-japan@wiley.com; Tel: +65 6511 8010 or Tel (toll-free): 005 316 50 480. Visit our Online Customer Get-Help available in 6 languages at www.wileycustomerhelp.com.

Information for Subscribers: *Annals of the New York Academy of Sciences* is published in 30 volumes per year. Subscription prices for 2013 are: Print & Online: US$6,053 (US), US$6,589 (Rest of World), €4,269 (Europe), £3,364 (UK). Prices are exclusive of tax. Australian GST, Canadian GST, and European VAT will be applied at the appropriate rates. For more information on current tax rates, please go to www.wileyonlinelibrary.com/tax-vat. The price includes online access to the current and all online back files to January 1, 2009, where available. For other pricing options, including access information and terms and conditions, please visit www.wileyonlinelibrary.com/access.

Delivery Terms and Legal Title: Where the subscription price includes print volumes and delivery is to the recipient's address, delivery terms are Delivered at Place (DAP); the recipient is responsible for paying any import duty or taxes. Title to all volumes transfers FOB our shipping point, freight prepaid. We will endeavour to fulfill claims for missing or damaged copies within six months of publication, within our reasonable discretion and subject to availability.

Back issues: Recent single volumes are available to institutions at the current single volume price from cs-journals@wiley.com. Earlier volumes may be obtained from Periodicals Service Company, 11 Main Street, Germantown, NY 12526, USA. Tel: +1 518 537 4700, Fax: +1 518 537 5899, Email: psc@periodicals.com. For submission instructions, subscription, and all other information visit: www.wileyonlinelibrary.com/journal/nyas.

Production Editors: Kelly McSweeney and Allie Struzik (email: nyas@wiley.com).

Commercial Reprints: Dan Nicholas (email: dnicholas@wiley.com).

Membership information: Members may order copies of *Annals* volumes directly from the Academy by visiting www. nyas.org/annals, emailing customerservice@nyas.org, faxing +1 212 298 3650, or calling 1 800 843 6927 (toll free in the USA), or +1 212 298 8640. For more information on becoming a member of the New York Academy of Sciences, please visit www.nyas.org/membership. Claims and inquiries on member orders should be directed to the Academy at email: membership@nyas.org or Tel: 1 800 843 6927 (toll free in the USA) or +1 212 298 8640.

Printed in the USA by The Sheridan Group.

View *Annals* online at www.wileyonlinelibrary.com/journal/nyas.

Abstracting and Indexing Services: *Annals of the New York Academy of Sciences* is indexed by MEDLINE, Science Citation Index, and SCOPUS. For a complete list of A&I services, please visit the journal homepage at www. wileyonlinelibrary.com/journal/nyas.

Access to *Annals* is available free online within institutions in the developing world through the AGORA initiative with the FAO, the HINARI initiative with the WHO, and the OARE initiative with UNEP. For information, visit www. aginternetwork.org, www.healthinternetwork.org, www.oarescience.org.

Annals of the New York Academy of Sciences accepts articles for Open Access publication. Please visit http://olabout.wiley.com/WileyCDA/Section/id-406241.html for further information about OnlineOpen.

Wiley's Corporate Citizenship initiative seeks to address the environmental, social, economic, and ethical challenges faced in our business and which are important to our diverse stakeholder groups. Since launching the initiative, we have focused on sharing our content with those in need, enhancing community philanthropy, reducing our carbon impact, creating global guidelines and best practices for paper use, establishing a vendor code of ethics, and engaging our colleagues and other stakeholders in our efforts. Follow our progress at www.wiley.com/go/citizenship.

Ann. N.Y. Acad. Sci. ISSN 0077-8923

ANNALS OF THE NEW YORK ACADEMY OF SCIENCES
Issue: *Inositol Phospholipid Signaling in Physiology and Disease*

Introduction to *Inositol Phospholipid Signaling in Physiology and Disease*

In June 2012, a two-day meeting was held at the New York Academy of Sciences in New York to discuss recent advances in the field of inositol phospholipid (IP) signaling. Over the last five to ten years, IP signaling has enjoyed something of a "great leap forward" because of the development of sophisticated mutant mouse models that have enabled the physiological role of key enzymatic components in this pathway to be revealed. In addition, small molecule inhibitors (or agonists) of these same enzymes have, in many cases, confirmed the genetic findings; these agonists have also importantly served as a means to preclinical explorations that have subsequently led to clinical evaluation of these agents. Indeed, some of these compounds are nearing federal approval as therapeutic agents—a major milestone for the IP signaling field. In the years to come we will likely see additional approvals that will increase the armaments available for treatment of oncologic, metabolic, and inflammatory diseases. This is particularly true of PI3K and its multiple isoforms. Although further back in the research and development pipeline, inhibitors and agonists for the SHIP1 and SHIP2 enzymes have also enabled similar preclinical studies; early human testing has begun in the case of SHIP1 agonists. The disease settings in which these IP enzymes appear to have prominent roles include cancer, inflammation, and metabolism. Below we provide a brief introduction to the papers in this volume, which offer insights into each of these disease settings.

Inositol phospholipid signaling in oncology: In the clinic and moving forward

Tyrosine kinase inhibitors (TKI) were among the first molecularly targeted therapies; in some instances they have revolutionized the treatment of disease and improved survival of cancer patients (e.g., in chronic myelogenous leukemia). Because these tyrosine kinases phosphorylate and activate PI3K, its subunits and various isoforms have become the next fertile area for research in cancer progression and persistence. In addition, the role that other inositol phosphatases—which can either limit or, in some cases, enhance PI3K signaling—play in malignancy has also been a major area of investigation over the last decade. This is particularly true for the $3'$-inositol phosphatase PTEN, which reverses the reaction catalyzed by PI3K isoforms. In this volume, Bertucci and Mitchell provide a succinct and eloquent summary of the impact of PI3K and PTEN mutations in breast cancer that serves as a prequel to their discussion of inositol 4-phosphatase B (INPP4B) mutations in breast cancer. Because INPP4B is now known to be a tumor suppressor, as shown by both Christina Mitchell's and Lewis Cantley's groups, and to degrade $PI(3,4)P_2$, it logically follows that the enzymes that produce $PI(3,4)P_2$, SHIP1, and SHIP2 might facilitate the growth of cancer. Indeed, as discussed in the article by Fernandes, Iyer, and Kerr, both SHIP1 and SHIP2 have been co-opted by cancer cells to facilitate their survival by limiting induction of both intrinsic and extrinsic cell death. These observations were enabled by the recent development by Kerr *et al.* of both SHIP1 selective and pan SHIP1/2 inhibitory compounds that are quite effective in killing of hematologic and breast cancers. These findings are a bit of a paradigm shift for the field, in that SHIP1/2, which also degrade $PI(3,4,5)P_3$, the PI3K product, were anticipated to be potent tumor suppressors. This may not be the

doi: 10.1111/nyas.12097

case in most instances, and particularly for chemical intervention against SHIP1 and 2. These findings speak to the importance of the cancer cell's requirement to sustain adequate levels of both $PI(3,4,5)P_3$ and $PI(3,4)P_2$, as summarized by Fernandes, Iyer, and Kerr in their "two PIP hypothesis" model. Although there is great hope for novel therapies that target IP signaling enzymes in cancer, Sung Su Yea and David A. Fruman offer us an illuminating discussion of the challenges the field faces regarding resistance to such therapeutic approaches. They initially provide a concise description of PI3K signaling and its activation of mTOR components that contribute to cancer cell survival, but they also discuss how the cancer cell may eventually become resistant to such approaches. They describe three models for how therapeutic resistance to PI3K/mTOR inhibition develops: compensatory changes by other IP signaling components when one specific component is targeted, epigenetic changes by the tumor cell, and mitochondrial priming that requires upregulation of BCL2 family members. They propose that the latter two resistance mechanisms could be avoided by simultaneously targeting a PI3K/mTOR component and HDAC enzymes or BCL2 members, respectively. Inasmuch as HDAC inhibitors (HDACi) and BCL2 agonists are already in the clinic, these combinatorial approaches would appear to be the next logical step in clinical testing of compounds that target PI3K and mTOR components in cancer. These types of combinatorial approaches were also strongly supported by keynote speaker Lewis C. Cantley (Beth Israel, Harvard Medical School), whose recent work has highlighted the possible favorable usage of metformin as a treatment "enhancer" for cancer (paper not included in this volume). Finally, Cristian Massacesi and colleagues offer the Novartis perspective to targeting of the IP signaling pathway in cancer, with a focus on PI3K and its isoforms. Inasmuch as Novartis pioneered the development and use of TKI in oncology, this use is a significant bellwether for targeting of the IP signaling components and that Novartis is also committed to exploring clinically this aspect of cancer signaling. Here Novartis's substantial and sophisticated experience in the development of TKI will likely prove invaluable to the field. This has already been revealed, as Massacesi *et al.* elaborate the importance of retrospectively stratifying responses in clinical trials of PI3K/mTOR inhibitors to better understand how underlying mutations in pathway components may have affected the response to therapy. Importantly, they also propose that prospectively characterizing and selecting patients on the basis of genetic mutations in IP signaling components and oncogenic loci will be critical to the successful application of novel IP signaling therapeutics in oncology.

Opportunities for novel therapeutic interventions in inflammatory diseases

A decade or more of *ex vivo* studies with broadly acting and relatively unselective PI3K inhibitors like wortmannin and LY294002 taught us that most or all of the adaptive and innate immune system functions that contribute to inflammation require IP signaling. However, revealing the physiological contribution of each IP signaling component required the development of sophisticated genetic models, particularly those utilizing Cre-lox–mediated inducible or cell type–specific ablation of IP signaling enzymes. In the papers devoted to inflammatory disease, the use of these models was essential in increasing our understanding of how these pathways contribute to inflammatory disease. However, as with cancer, where genetics preceded and fostered pharmacologic intervention, inflammatory diseases are now just beginning to be targeted with chemical entities showing greater selectivity for IP signaling components than wortmannin and Ly294002. Of particular value in this decade-long effort was work by Klaus Okkenhaug and colleagues who systemically defined the contribution of the four different PI3K class I isoforms (p110α, β, γ, and δ) to immunity and inflammation using genetic models that selectively target each of these isoforms. Okkenhaug presents a concise summary of their role in promoting the effector functions of B lymphocytes, T lymphocytes, and neutrophils, and proposes how dual inhibition of the four PI3K p110 isoforms may see clinical application in autoimmunity and inflammation.

The paper by Huynh and Turka serves as an important contrast by highlighting the pivotal role that the enzyme that opposes the above four PI3K isoforms plays in limiting T_{reg} cell development and function. This phenomenon illustrates the complexities of targeting PI3K components, as PTEN- and SHIP1-mutant mice revealed that PI(3,4,5)P3 is crucial not only in the physiological responses that lead to inflammation, but also in its control by immunoregulatory T_{reg} cells. Further adding to the complexities surrounding the feasibility of targeting PI3K isoforms in inflammation is the demonstration that the potent inducers of mucosal inflammation, Th17 cells, also require PI3K for their generation and function *in vivo*, as described by Nagai and colleagues. Fernandes, Iyer, and Kerr take the discussion of mucosal inflammation and IP signaling further down the pathway by summarizing their studies that surprisingly show that selective T cell deficiency at mucosal surfaces leads to lung and GI tract pathologies in SHIP-mutant mice. Sriskantharajah and colleagues from GlaxoSmithKline make a strong case in their paper for the feasibility of targeting the PI3K p110δ isoform in lung inflammatory diseases and particularly in chronic obstructive pulmonary disease. Wai-Ping Fung-Leung from Janssen Research & Development makes an equally compelling case for targeting of the PI3K p110γ isoform in inflammatory and autoimmune diseases. In addition, she proposes that dual inhibitors, rather than selective targeting of the δ or γ p110 isoforms, might be more clinically efficacious. Although less developed clinically than in oncology, small molecule targeting of IP signaling components appears to have strong therapeutic potential using murine genetic studies and, in some cases, results from early clinical testing with various PI3K inhibitors.

Emerging opportunities for therapeutic application to metabolic disease

An emerging aspect of IP signaling can be seen in its role in metabolism that spans thermogenesis, insulin signaling, obesity, and bone biology. As elaborated by Wymann and Solinas, the highly complex physiological role of IP signaling components may not be unrelated to their role in inflammation, particularly for macrophage populations residing in tissues and in the cardiovascular system. Here they make a case for specifically targeting PI3K p110γ in both inflammation and cardiovascular disease. Jaber and Wei-Xing Zong discuss the fundamental role that the class III PI3K, Vps34, plays in autophagy and how this role contributes to heart and liver function. Understanding the function of this primitive PI3K enzyme, which is conserved in such simple organisms as yeast, may yield unanticipated therapeutic possibilities. Jean Vacher, and Iyer, Margulies, and Kerr provide a discussion of the pivotal role that inositol phosphatases INPP4B and SHIP1 play in bone development and homeostasis, respectively. Vacher and colleagues have revealed a fundamental role for INNP4B in restraining osteoclast function *in vivo* using the INPP4B$^{-/-}$ mice they developed. Their findings provide further insight into key mechanisms triggered by IP signaling in the immune system as it pertains to systemic physiology, echoing a similar proposal by Wymann and Solinas for cardiovascular disease and obesity. By contrast, Iyer, Margulies, and Kerr show that IP signaling in mesenchymal stem cells and osteoprogenitors has an equally important role in bone physiology through the development and study of mice where SHIP1 expression is selectively ablated in these mesenchymal lineage cells. Moreover, the potential of these findings for the metabolic control of body fat and muscle mass by osteolineage cells, and through interventions like treatment with SHIP1 inhibitors, is revealed by their genetic studies. These observations once again highlight the ongoing contributions that sophisticated murine genetic models are making to our understanding of whether and how IP signaling components contribute to normal physiology and disease, and how we might target this pathway in disease.

Overall, a substantial increase in our understanding of the mechanisms of IP signaling in physiology and disease has served as a major catalyst for the development of novel selective therapeutics targeting this signaling pathway over the last decade. Many challenges, however, still lie ahead. Indeed, in his plenary presentation (not included in this volume) Neal Rosen (Memorial Sloan-Kettering Cancer Center) pointed out that successful manipulation of the IP signaling cascade will undoubtedly require a more profound understanding of the balance regulating the entire network of inositol 5′, 4′, and 3′ phosphatases and kinases, such that this field will continue to spark interesting debates and controversies in the coming years.

WILLIAM G. KERR[1,2] AND SANDRA FERNANDES[1]

[1]*Departments of Microbiology and Immunology, and* [2]*Pediatrics*
State University of New York, Upstate Medical University
Syracuse, New York

Ann. N.Y. Acad. Sci. ISSN 0077-8923

ANNALS OF THE NEW YORK ACADEMY OF SCIENCES
Issue: *Inositol Phospholipid Signaling in Physiology and Disease*

Phosphoinositide 3-kinase and INPP4B in human breast cancer

Micka C. Bertucci and Christina A. Mitchell

Department of Biochemistry and Molecular Biology, Monash University, Clayton, Victoria, Australia

Address for correspondence: Christina A. Mitchell, Department of Biochemistry and Molecular Biology, Monash University, Clayton, Victoria 3800, Australia. Christina.Mitchell@monash.edu

The PI3K/Akt signaling pathway is frequently increased in many human cancers, including breast cancer. Recent studies have identified INPP4B, which inhibits PI3K signaling, as an emerging tumor suppressor in breast cancer. This short review discusses these issues and the possibility that INPP4B is an important regulator in many cancers.

Keywords: PI3K; breast cancer; PTEN; 4-phosphatase; INPP4B

Introduction

In response to extracellular stimuli, phospho-inositide 3-kinase (PI3K) is activated, resulting in the phosphorylation of membrane-bound phosphatidylinositol 4,5-bisphosphate $(PtdIns(4,5)P_2)$ to transiently generate phosphatidylinositol 3,4,5-trisphosphate $(PtdIns(3,4,5)P_3)$ that binds the pleckstrin homology (PH) domain of many proteins, including phosphoinositide-dependent kinase 1 (PDK1) and the proto-oncoprotein serine/threonine kinase Akt.[1] Phosphoinositide binding to these PI3K effectors leads to their tyrosine phosphorylation and/or allosteric activation and, in turn, the activation of other downstream targets, including the mammalian target of rapamycin (mTOR) complex, which promotes cell growth and protein translation; serum glucocorticoid regulated kinase (SGK); and other signaling pathways that promote cell cycle progression and cell survival. $PtdIns(3,4,5)P_3$ is rapidly dephosphorylated by inositol polyphosphate 5-phosphatases leading to the production of a second signaling lipid $PtdIns(3,4)P_2$. This phosphoinositide also binds the PH domain of Akt with high affinity and, with $PtdIns(3,4,5)P_3$, facilitates full activation of Akt.[1,15] PI3K signaling is negatively regulated by several phosphoinositide phosphatases, including the tumor suppressor PTEN, which dephosphorylates the 3-position phosphate from the inositol ring

of $PtdIns(3,4,5)P_3$ to form $PtdIns(4,5)P_2$. More recently a novel tumor suppressor INPP4B has been identified, which dephosphorylates the 4-position phosphate from the inositol ring of $PtdIns(3,4)P_2$ to form $PtdIns(3)P$, which also inhibits PI3K-dependent Akt activation. Enhanced PI3K/Akt signaling has been identified in many human cancers. In primary human breast cancer, around 70% of cases exhibit alterations within one or more components of the PI3K pathway,[2] including mutation or amplification of the gene encoding the catalytic subunit of class I PI3K (*PIK3CA*); activation of oncogenic tyrosine kinase receptors such as HER2, which, in turn, activate PI3K/Akt; loss of function of the tumor suppressors PTEN and/or INPP4B; or mutation and/or amplification of the proto-oncogene *AKT1* (Fig. 1). The focus of this review is the role of PI3K signaling and its regulation by INPP4B in human breast cancer.

Phosphoinositide 3-kinase mutations in breast cancer

Invasive breast cancer is a heterogeneous disease that is generally classified as distinct subtypes—luminal A, luminal B, HER2-positive, and triple-negative—on the basis of the expression of specific receptors and clinicopathological markers.[12] Luminal A and B breast cancers are hormone dependent and express the estrogen receptor (ER) alpha and/or progesterone receptor (PR), and are responsive to

doi: 10.1111/nyas.12036

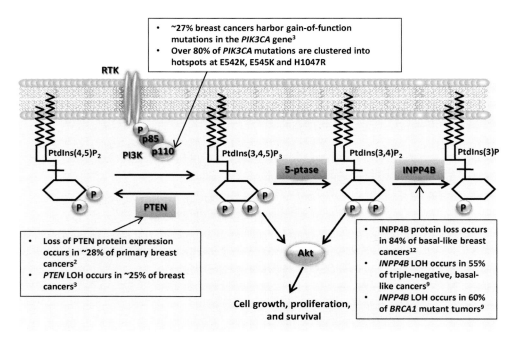

Figure 1. The PI3K signaling pathway. Activation of PI3K at the plasma membrane by receptor tyrosine kinase (RTK) results in the transient production of phosphatidylinositol-3,4,5-trisphosphate (PtdIns(3,4,5)P$_3$) that is hydrolyzed by the inositol polyphosphate 5-phosphatases (SHIP1/2) to generate PtdIns(3,4)P$_2$. Both PtdIns(3,4,5)P$_3$ and PtdIns(3,4)P$_2$ facilitate the activation of Akt. PI3K signaling can be terminated through the actions of the 3-phosphatase, PTEN, which hydrolyzes PtdIns(3,4,5)P$_3$ to produce PtdIns(4,5)P$_2$. PtdIns(3,4)P$_2$ can be hydrolyzed by the actions of the 4-phosphatase, INPP4B, to generate PtdIns(3)P. Alterations that have been reported in these enzymes in breast cancer are indicated in the boxed areas.

treatments that block ER signaling. HER2-positive breast cancers exhibit amplification or overexpression of the proto-oncogene *ERBB2* (HER2) and respond to blockade of HER2 receptor signaling. Triple-negative breast cancers do not express ER, PR, or overexpress HER2, and for this subtype, chemotherapy is the main treatment option; in general, these cancers exhibit a worse prognosis. Basal-like breast cancers can be distinguished from triple-negative cancers through the assessment of expression of basal markers such as cytokeratin 5/6 and epidermal growth factor receptor (EGFR).[12]

It is estimated that approximately 27% of human breast tumors carry gain-of-function mutations in the *PIK3CA* gene (as reviewed in Liu *et al.*).[3] Class I PI3K family members are heterodimeric lipid kinases comprising catalytic p110 and regulatory p85 subunits. Somatic missense mutations in *PIK3CA* are clustered into two hotspot regions in exons 9 and 20, corresponding to the helical and kinase domains of p110α, respectively. More than 80% of PIK3CA mutations occur at E542K, E545K, or H1047R, and these mutations are oncogenic, re-

sulting in increased Akt activation.[3] *PIK3CA* mutations are most frequently observed in hormone receptor-positive breast tumors and HER2-positive tumors but rarely in the triple-negative/basal-like cancers.[4] In some studies *PIK3CA* mutations occur in a mutually exclusive manner from PTEN loss of function.[2] However, concordance of *PIK3CA* mutational events with PTEN loss in breast cancer has also been described.[4] The association between *PIK3CA* mutations and long term survival is complex, some studies have shown a surprising correlation between *PIK3CA* mutations and good prognosis.[5] Gene expression and protein data from about 1800 breast cancers revealed that *PIK3CA* mutations in ER-positive/HER2-negative breast cancers were associated with Akt pathway activation but a relatively low mTORC1 signaling profile, and predict better clinical outcomes with tamoxifen (which blocks ER signaling) therapy.[5] Mouse models that express the PIK3CA-H1047R mutant in luminal mammary epithelium recapitulate the heterogeneity of breast cancer, with induction of carcinomas containing cells that express luminal or basal markers, or both, with ER expression also observed.[6] Liu *et al.* also generated transgenic mice expressing mutant human PIK3CA-H1047R under the control of a tetracycline-inducible promoter in mammary tissue. Heterogeneous breast cancer occurred in 95% of mice within seven months, and tumors regressed following switching off the transgene, but this regression was not complete and most tumors reoccurred, associated with Myc amplification.[7] Therefore mutations of PIK3CA that lead to activation of Akt signaling are common in ER-positive breast cancer, and appear to initiate breast cancer in mouse models; however their prognostic significance is still emerging.

Role of PTEN in breast cancer

PTEN is a tumor suppressor gene located at 10q23, a region frequently mutated in human cancers. PTEN is a negative regulator of PI3K activity that hydrolyzes PtdIns(3,4,5)P$_3$ to PtdIns(4,5)P$_2$.[1,3] PTEN contains an amino-terminal phosphatase domain with a conserved catalytic CX$_5$R motif and a carboxyl-terminal domain containing a C2 domain. Loss of PTEN function increases PtdIns(3,4,5)P$_3$ signals and leads to unrestrained Akt activation and increased cell survival, growth, and proliferation.[3]

In cancer, reduced PTEN function can occur via mutation, loss of heterozygosity (LOH), protein instability, and/or epigenetic modifications. Germline mutations of *PTEN* are present in 80% of families with autosomal dominant Cowden's syndrome, which is associated with an elevated risk of breast cancer. Somatic and biallelic mutations of *PTEN* are common in high grade glioblastoma and melanoma, as well as prostate and endometrial cancers; however, *PTEN* mutations in sporadic breast cancer are less frequent at 2.3% incidence and appear limited to ER/PR-positive tumors.[4] LOH at the *PTEN* locus, 10q23, is found in 24.9% of breast carcinomas.[3] PTEN protein expression is lost in approximately 28% of primary breast carcinomas.[2] Loss of PTEN protein expression may be associated with enhanced lymph node metastasis and disease-related death. However, in contrast, a study of 292 breast cancer patients found no association between low PTEN levels and metastasis or disease-related death, but observed a correlation with the poor prognosis basal-like breast cancer subtype.[2] *Pten*-null mice exhibit early embryonic lethality due to developmental defects. *Pten*$^{+/-}$ mice

develop spontaneous tumors of the thyroid, colon, and gonado-stromal tissues. Notably, a *Pten* hypermorphic mouse, *Pten*$^{hy/+}$, has been generated that expresses 80% of wild-type levels of Pten. These mice spontaneously develop a range of tumors, with mammary tumors at the highest prevalence.[8] Significantly, the spectrum of tumors exhibited by *Pten*$^{hy/+}$ mice are distinct from those observed with *Pten*$^{+/-}$ mice, suggesting that subtle alterations in Pten expression may promote tumorigenesis in a tissue-specific manner.

The role of the inositol polyphosphate 4-phosphatase INPP4B in breast cancer

Recent studies have identified that the inositol polyphosphate 4-phosphatase, INPP4B, may function as a tumor suppressor in breast cancer. INPP4B, and the related INPP4A, hydrolyze PtdIns(3,4)P$_2$, producing PtdIns(3)P, and also display catalytic activity toward the inositol phosphates Ins(1,3,4)P$_3$ and Ins(3,4)P$_2$. However, PtdIns(3,4)P$_2$ is the preferred *in vivo* INPP4A/B substrate.[9] INPP4A and INPP4B share 37% amino acid identity, with N-terminal C2 domains and a C-terminal catalytic CX$_5$R motif. INPP4A is critical for normal neuronal function. A spontaneous *Inpp4a* mutation in the mouse strain weeble (*Inpp4a*^{wbl}) results from a single base pair deletion (Δ744). The weeble mouse dies two to three weeks postnatally, with cerebellar ataxia and growth retardation.[10] Gene targeted deletion of *Inpp4a* in mice leads to a similar phenotype, with evidence of glutamate neuroexcitatory cell death. INPP4A also regulates PI3K/Akt-dependent signaling and cell survival. Weeble mouse embryonic fibroblasts (MEFs) exhibit Akt activation and anchorage-independent cell growth.[11] SV40-transformed weeble MEFs form tumors in nude mice.[11] However, there is limited evidence of changes in INPP4A expression in human cancers.

There is emerging evidence that INPP4B functions as a putative tumor suppressor in several human cancers, including breast cancer, prostate cancer, and melanoma. INPP4B regulates PI3K/Akt signaling in breast cancer cell lines. Small hairpin RNA (shRNA)-mediated INPP4B knockdown in ER-positive MCF-7 breast cancer cells results in increased Akt activation, with enhanced colony formation in soft agar and increased xenograft tumor growth in nude mice.[12] INPP4B overexpression in SUM149 cells, a *BRCA1* mutated human invasive ductal carcinoma cell line, decreases tumor growth in a xenograft mouse model.[9] Knockdown of both INPP4B and PTEN in human mammary epithelial cells induces cell senescence, a phenotype rescued by p53 depletion.[9] In normal breast, INPP4B protein expression is limited to ER-positive mammary ductal luminal epithelial cells.[12] Consistent with this, INPP4B protein expression is restricted to ER-positive breast cancer cell lines.[12] However, INPP4B expression does not appear to be regulated by ER signaling. In human breast cancers, INPP4B expression is strongly associated with both ER and PR expression. Notably, immunohistochemical INPP4B antibody staining of 374 primary human breast carcinomas revealed frequent loss of INPP4B protein expression in aggressive basal-like breast carcinomas (84% of cases).[12] LOH at 4q31.21, the chromosomal region containing *INPP4B*, is also reported to occur in triple-negative/basal-like tumors (55% of cases).[9] The lower frequency for LOH, compared to loss of protein expression of INPP4B in triple-negative carcinomas, suggests differences in the rates of allelic and protein INPP4B loss. *INPP4B* LOH is most commonly detected in *BRCA1* mutant tumors (60% of cases).[9] *BRCA1* mutant and basal-like carcinomas are typically triple-negative for ER/PR and HER2 expression. *INPP4B* LOH also occurs in ovarian cancers (39.8% of cases).[9] Reduced INPP4B protein expression correlates with decreased overall patient survival for both breast and ovarian cancer.[9] The triple-negative/basal-like breast tumor subtype, in which *INPP4B* LOH and decreased protein expression is observed, is typically more metastatic with an unfavorable patient outcome. In a breast carcinoma patient cohort with few *BRCA1* mutant and basal-like subclasses, loss of INPP4B protein expression significantly correlated with reduced overall patient survival.[9] Therefore, INPP4B expression in breast cancer may play a role in predicting long term survival. As loss of INPP4B is associated with enhanced PI3K-dependent Akt signaling in both cell lines and tissues, it is likely these tumors may respond to treatment with PI3K and other pathway inhibitors. No association has been reported between INPP4B protein expression and *PIK3CA* mutation or amplification. However, in one report INPP4B protein loss was significantly associated with PTEN loss of function and high pAkt levels. In constrast, loss of INPP4B protein

expression by itself is not associated with high pAkt levels.[12] Both PtdIns(3,4)P$_2$ and PtdIns(3,4,5)P$_3$ are required for full and sustained activation of Akt.[1] Therefore, concurrent loss of INPP4B and PTEN, which degrade PtdIns(3,4)P$_2$ and PtdIns(3,4,5)P$_3$, respectively, may lead to the development of more aggressive forms of breast cancer with Akt hyperactivation. Consistent with the role of PtdIns(3,4)P$_2$ and INPP4B in breast cancer cell survival, SHIP2 inhibitory compounds that block PtdIns(3,4)P$_2$ production are quite effective at killing breast cancer cells.[16]

INPP4B is also implicated as a tumor suppressor in prostate cancer, though with only one study to date. In primary prostate tumors, INPP4B and PTEN show reduced expression compared with normal tissue; decreased expression of INPP4B is associated with reduced time to biochemical evidence of prostate cancer recurrence.[13] *INPP4B* LOH has also been reported in a high proportion of melanomas (21.6% of cases).[9]

Recently, the first *Inpp4b* knockout mouse was generated and characterized. Inpp4b is a significant regulator of osteoclast differentiation that can have a significant impact on bone mass.[14] *Inpp4b*-deficient mice display osteoporosis, due to increased osteoclast cell populations and enhanced bone resorption.[14] Interestingly, ablation of *Inpp4b* did not result in spontaneous tumor development in mice up to four months of age. Therefore, INPP4B loss alone may not be sufficient to cause tumor development, with additional PI3K pathway aberrations required. Alternatively, INPP4B loss of function may be critical at later stages of tumor development.

Summary

Recent studies have identified INPP4B as a novel tumor suppressor that inhibits PI3K/Akt signaling and proliferation in ER-positive mammary cancer cells. Loss of INPP4B expression has been observed predominantly in basal-like/triple-negative breast cancers and may be an additional predictive marker of poor outcome. Furthermore, there is evidence that INPP4B is concurrently lost with PTEN in these poor prognosis breast cancer subtypes, and it is likely these tumors may be candidates for treatment with PI3K pathway inhibitors. In addition, several recent studies suggest INPP4B expression may be altered in melanomas and prostate cancer. Therefore it is likely that this newly identified tumor suppressor gene may be an important regulator of PI3K/Akt signaling in many human cancers.

Acknowledgments

This work was supported by a Grant from the National Health and Medical Research Council (606621) (CAM).

Conflicts of interest

The authors declare no conflicts of interest.

References

1. Ma, K. *et al.* 2008. PI(3,4,5)P$_3$ and PI(3,4)P$_2$ levels correlate with PKB/Akt phosphorylation at Thr308 and Ser473, respectively; PI(3,4)P$_2$ levels determine PKB activity. *Cell. Signal.* **20:** 684–694.

2. López-Knowles, E. *et al.* 2010. PI3K pathway activation in breast cancer is associated with the basal-like phenotype and cancer-specific mortality. *Int. J. Cancer* **126:** 1121–1131.

3. Liu, P. *et al.* 2009. Targeting the phosphoinositide 3-kinase pathway in cancer. *Nat. Rev. Drug Discov.* **8:** 627–644.

4. Stemke-Hale, K. *et al.* 2008. An integrative genomic and proteomic analysis of *PIK3CA*, *PTEN*, and *AKT* mutations in breast cancer. *Cancer Res.* **68:** 6084–6091.

5. Loi, S. *et al.* 2010. *PIK3CA* mutations associated with gene signature of low mTORC1 signaling and better outcomes in estrogen receptor–positive breast cancer. *Proc. Natl. Acad. Sci.* **107:** 10208–10213.

6. Meyer, D. S. *et al.* 2011. Luminal expression of *PIK3CA* mutant H1047R in the mammary gland induces heterogeneous tumors. *Cancer Res.* **71:** 4344–4351.

7. Liu, P. *et al.* 2011. Oncogenic PIK3CA-driven mammary tumors frequently recur via PI3K pathway-dependent and PI3K pathway-independent mechanisms. *Nat. Med.* **17:** 1116–1120.

8. Alimonti, A. *et al.* 2010. Subtle variations in Pten dose determine cancer susceptibility. *Nat. Genet.* **42:** 454–458.

9. Gewinner, C. *et al.* 2009. Evidence that inositol polyphosphate 4-phosphatase type II is a tumor suppressor that inhibits PI3K signaling. *Cancer Cell* **16:** 115–125.

10. Nystuen, A. *et al.* 2001. A null mutation in inositol polyphosphate 4-phosphatase type I causes selective neuronal loss in Weeble mutant mice. *Neuron* **32:** 203–212.

11. Ivetac, I. *et al.* 2009. Regulation of PI(3)K/Akt signaling and cellular transformation by inositol polyphosphate 4-phosphatase-1. *EMBO Rep.* **10:** 487–493.

12. Fedele, C.G. *et al.* 2010. Inositol polyphosphate 4-phosphatase II regulates PI3K/Akt signaling and is lost in human basal-like breast cancers. *Proc. Natl. Acad. Sci.* **107:** 22231–22236.

13. Hodgson, M.C. *et al.* 2011. Decreased expression and androgen regulation of the tumor suppressor gene INPP4B in prostate cancer. *Cancer Res.* **71:** 572–582.

14. Ferron, M. *et al.* 2011. Inositol Polyphosphate 4-phosphatase B as a regulator of bone mass in mice and humans. *Cell Metabol.* **14:** 466–477.

15. Franke, T.F. *et al.* 1997. Direct regulation of the Akt proto-oncogene product by phosphatidylinositol-3,4-bisphosphate. *Science* **275:** 665–668.

16. Fuhler G.M. *et al.* 2012. Therapeutic potential of SH2 domain-containing inositol-5'-phosphotase 1 (SHIP1) and SHIP2 inhibition in cancer. *Mol. Med.* **18:** 65–75.

Ann. N.Y. Acad. Sci. ISSN 0077-8923

ANNALS OF THE NEW YORK ACADEMY OF SCIENCES

Issue: *Inositol Phospholipid Signaling in Physiology and Disease*

Role of SHIP1 in cancer and mucosal inflammation

Sandra Fernandes,[1] Sonia Iyer,[1] and William G. Kerr[1,2]

[1]Department of Microbiology and Immunology, SUNY Upstate Medical University, Syracuse, New York. [2]Department of Pediatrics, SUNY Upstate Medical University, Syracuse, New York

Address for correspondence: William G. Kerr, Ph.D., SUNY Upstate Medical University, 750 E. Adams Street, 2204 Weiskotten Hall, Syracuse, NY 13210. kerrw@upstate.edu

The SH-2 containing inositol 5′-polyphosphatase 1 (SHIP1) is a multifunctional protein expressed predominantly, but not exclusively, by hematopoietic cells. SHIP1 removes the 5′-phosphate from the product of PI3K, $PI(3,4,5)P_3$, to generate $PI(3,4)P_2$. Both PIP species influence the activity level of Akt and ultimately regulate cell survival and differentiation. SHIP1 also harbors several protein interaction domains that endow it with many nonenzymatic cell signaling or receptor masking functions. In this review, we discuss the opposing roles of SHIP1 in cancer and in mucosal inflammation. On one hand, germline loss of SHIP1 causes myeloid lung consolidation and severe inflammation in the ileum, a phenotype that closely mimics human Crohn's disease and can be rescued by reconstitution with SHIP1-competent T cells. On the other, transient inhibition of the enzymatic activity of SHIP1 in cancer cells leads to apoptosis and enhances survival in lethal murine xenograft models. Overall, careful dissection of the different pathological mechanisms involved in several diseases provides novel opportunities for therapeutic intervention targeting SHIP1.

Keywords: SHIP1; cancer; Crohn's disease; T cell; caspase 8

Introduction

The SH2-containing inositol 5′-polyphosphatase (SHIP1) is expressed predominantly by cells in the hematopoietic compartment,[1] but also by osteoblasts,[2] and it is encoded by the INPP5D gene. The highly conserved enzymatic domain is centrally located within the protein and is flanked on the N-terminal side by a PH-like domain that binds the SHIP substrate phosphatidylinositol 3, 4, 5-triphosphate $(PI(3,4,5)P_3)$,[3] and on the C-terminal side by a C2-domain that binds the product phosphatidylinositol 3,4 bisphosphate $(PI(3,4)P_2)$[4] (Fig. 1A). By virtue of its dephosphorylation of the product of phosphatidylinositol 3-kinase (PI3K), SHIP is a key player in the inositol phospholipid signaling cascade that also involves the tumor suppressor PTEN, which reverses the PI3K reaction by removing the 3′-phosphate of $PI(3,4,5)P_3$, and the INPP4A/B enzymes, which dephosphorylate the SHIP product, $PI(3,4)P_2$, at the 4′ position to generate $PI(3)P$ (Fig. 1B).[1] Moreover, nonenzymatic roles have been attributed to SHIP's N-terminal SH2 domain and its C-terminal NPYX

motifs and proline-rich regions. These C-terminal motifs are involved in interactions with proteins carrying phosphotyrosine-binding motifs (PTB), and for binding by SH3-domain containing proteins, respectively. SHIP can also be phosphorylated by cAMP-dependent PKA (Ser440, Fig. 1A), and this increases its enzyme activity;[5] however, the physiological role of this regulation remains to be determined. SHIP can also associate with receptor tails through its SH2 domain and mask key recruitment sites for other enzymes such as SHP1 or PI3K. Thus, SHIP1 can influence cell signaling in a manner that is independent of its enzymatic activity (Fig. 1C).[6,7] Because of its modular design, SHIP1 can have varied and disparate effects on cell signaling.

SHIP in cancer: the "two PIP hypothesis" of malignancy

Owing to their ability to reduce $PI(3,4,5)P_3$ levels at the plasma membrane, SHIP1, SHIP2 and PTEN are typically viewed as opposing the activity of the PI3K/Akt/mammalian target of rapamycin (mTOR) signaling axis that promotes cancer cell

doi: 10.1111/nyas.12038

Ann. N.Y. Acad. Sci. 1280 (2013) 6–10 © 2013 New York Academy of Sciences.

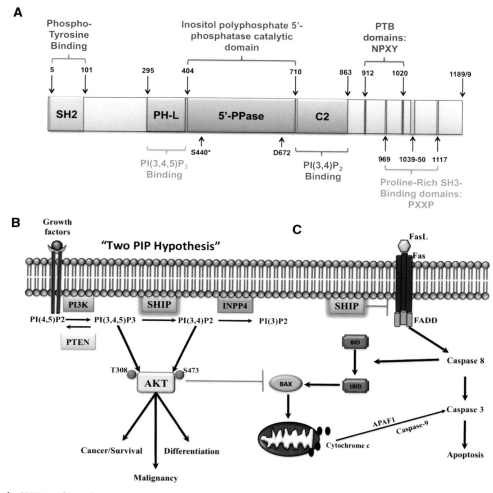

Figure 1. SHIP1 and its role in cancer and mucosal inflammation. (A) SHIP1 is a 1189 amino acid (1188 aa for isoform b) protein consisting of several different domains, namely the central catalytic domain (5′-PPase) flanked by a PH-like (PH-L) and a C2 domain for binding of the substrate PI(3,4,5)P_3 and product PI(3,4)P_2 of the enzyme, respectively. The N-terminal region contains a SH2-domain, while the C-terminal proline-rich region contains PXXP motifs for interactions with SH3-containing proteins and two NPXY motifs for interaction with phosphotyrosine binding (PTB) protein. Phosphorylation of Ser440 by PKA has been shown to increase enzymatic activity, and the aspartic acid at position 672 (D672) has been identified as a critical catalytic residue. (B) The two PIP hypothesis suggests that both the substrate, PI(3,4,5)P_3, and the product, PI(3,4)P_2, of SHIP are necessary to fully activate Akt in order to achieve and maintain the malignant state. As such, selective SHIP1 or pan-SHIP1/2 inhibition can trigger cancer cell death both *in vitro* and *in vivo*. (C) SHIP may also be involved in regulating Fas trimerization at the cell membrane and would thus serve a protective function against downstream caspase 8 signaling. Selective SHIP inhibition may therefore have a dual role in triggering apoptosis in cancer cells, via reduced AKT activation and increased caspase 8 activation. The latter mechanism also suggests that SHIP inhibition may prove useful in reducing inappropriate persistence of autoreactive effector T cells at mucosal surfaces by lifting the inhibition imposed on pathways downstream of Fas.

survival. However, emerging evidence suggests that in certain contexts, SHIP1 and SHIP2 may facilitate, rather than suppress, tumor cell survival as PTEN does.[8,9] Indeed, while PTEN and the SHIP1/2 proteins dephosphorylate the direct product of PI3K, PTEN removes the phosphate from the D3 position of the inositol ring, while SHIP1/2 remove

the D5 phosphate. This distinction is crucial, as it enables SHIP1/2 and PTEN to have very different effects on Akt signaling. Consistent with this hypothesis, PI(3,4)P_2 levels are increased in leukemic cells, and increased levels of PI(3,4)P_2 in *Inpp4b* (type II) mutant mice promote mammary epithelial cell transformation and tumorigenicity.[10] This phenomenon can be explained by the "two PIP hypothesis",[1] which states that a certain amount of both $PI(3,4,5)P_3$ and $PI(3,4)P_2$ are required to promote and maintain the malignant state (Fig. 1B). This hypothesis is further supported by the fact that both SHIP1 agonistic[4] and antagonistic[8] compounds have been shown to kill multiple myeloma (MM) cells.

We have also shown that a SHIP1-selective small molecule inhibitor, 3α-aminocholestane (3AC), reduced the viability of several acute myeloid leukemia (AML) cell lines, including KG-1 and C1498, in a dose-dependent manner while having no effect on leukemia cells that do not express SHIP1 such as K562.[8] Hallmarks of apoptosis, such as cleavage of PARP, caspase 3 and 9, and increased annexin V staining were observed following treatment with 3AC, indicating that cell death pathways were triggered by 3AC. Recently, we and others have observed that SHIP1 may prevent oligomerization of Fas,[11] and thus SHIP1 inhibition may help increase caspase 8 activation and promote apoptosis in cancer cells (Fig. 1B) (Sudan, Fernandes, Srivastava, Kerr, unpublished data). In addition, SHIP1 inhibition, via 3AC treatment, was shown to abrogate MM growth *in vivo* and enhance survival in a lethal xenograft model of MM in immunodeficient NOD/SCID/IL-2RγC (NSG) mice.[9] Significant reductions were observed in the tumor burden in 3AC-treated versus vehicle control mice, as measured by reduction of both human free lambda light chain in the serum and percentage of circulating MM cells in the peripheral blood. Most importantly, pulsatile treatment with the 3AC SHIP1 inhibitor for several months failed to trigger the pathology or morbidity that is universally observed in mice with germline loss of SHIP1 expression, including the lethal pulmonary pneumonia[12] and Crohn's-like ileitis[13] that afflict germline $Ship1^{-/-}$ mice.

More recently, we demonstrated that pan-SHIP1/2–targeted inhibition might prove useful for treatment of various blood cancers, such as MM and AML, as well as nonhematopoietic malignant cells that solely express SHIP2. Indeed, a dose-dependent decrease in viability was observed for both MCF-7 and MBA-MB-231 breast cancer cells following treatment with different pan-SHIP1/2 inhibitors.[9] As observed for SHIP1-selective inhibition, pan-SHIP1/2 inhibition reduced basal and IGF-1–induced Akt activation, suggesting that tar-

geting the SHIP1/2 proteins does indeed influence the Akt pathway downstream of PI3K (Fig. 1B). Overall, we believe that selective and/or pan-SHIP1/2–inhibition may be a promising novel therapeutic approach to counter the PI3K/Akt/mTOR pathway that is frequently mutated in cancer and is critical for survival of malignant cells.

Role of SHIP in inflammation at mucosal surfaces

In the first paper describing germline *Ship* mutant mice, Helgason *et al.* noted a profound myeloid consolidation of the lungs that they proposed led to the early demise of the mice.[12] We subsequently showed that this lethal disease could be cured by an allogeneic BM transplant from a SHIP-competent donor, indicating that the pulmonary disease resulted from a defect in cellular processes confined to the hematolymphoid compartment and under the control of SHIP1.[14] Consistent with this, we could recapitulate pulmonary consolidation in adult mice following induction of SHIP deficiency using an inducible *Ship1* deletion model (MxCre*Ship*$^{flox/flox}$ mice), demonstrating SHIP1 plays this role in the hematopoietic system throughout life, not only during neonatal or juvenile life.[8] The prevailing model to explain this lethal pneumonia is that it represented a myeloproliferative disease that affects alveolar spaces due to decreased turnover of SHIP1 deficient myeloid subsets.[12,15] However, this hypothesis was not directly tested *in vivo*. Subsequently we found that myeloid inflammatory disease also occurs in the small intestines of $Ship1^{-/-}$ mice,[13] including two strains in which *Ship* was mutated by targeting the promoter and first exon[12,16] and a third strain that lacked the inositol phosphatase exons of SHIP1.[17] This disease was highly demarcated, as it was confined to the terminal portion of the ileum where lymphoid tissue is abundant in the intestine; it was not found in the proximal portion of the small intestine, large intestine, or bowel.[13] Thus, in most of its aspects, the phenotype strongly resembles classic Crohn's disease (CD), as seen in human inflammatory bowel disease (IBD), and in particular those cases where inflammation has not propagated to the descending colon. Like the pulmonary inflammation seen in $Ship^{-/-}$ mice, histopathology and flow cytometry analysis of the small intestine that indicated a profound inflammatory myeloid infiltrate was present in the terminal ileum, with a strong neutrophil component.[13] However, a more intriguing observation was that both CD4[+] and CD8[+] T cell numbers in the small intestine lamina propria were greatly diminished in *Ship1*[−/−] mice,[13] suggesting that *Ship1*[−/−] mice might have a selective deficiency of effector T cells at mucosal sites. T cells are thought to play prominent roles in surveillance and immune responses to both commensal organisms and pathogens in the gut and lung. Hence, we proposed that the myeloid inflammatory disease in these mice results from a failure of this T cell function and consequently an overresponse by SHIP1-deficient myeloid cells that also provide pathogen surveillance and protection at mucosal sites.[13]

To test the above hypothesis, we selectively reconstituted sublethally irradiated *Ship*[−/−] mice with SHIP1-competent, wild-type T cell grafts. We found that if the T cell graft is given before 40 days of life, it is sufficient to rescue *Ship1*[−/−] mice from both pulmonary and gut inflammatory disease; in fact, it endowed them a normal and healthful lifespan. These findings indicated that SHIP1 plays a role in promoting the persistence of mature T cells in the periphery and at mucosal surfaces. We have confirmed this using both a small molecule inhibitor of SHIP1 and competition studies of wild-type and *Ship1*[−/−] T cells in both immunocompetent and SCID hosts that lack an endogenous T cell compartment. These studies revealed that SHIP1 promotes the persistence of T cells in the recirculating T cell compartment of the blood and spleen, as well as in the more specialized mucosal T cell compartments of the lung and small intestine. Our results do not rule out roles for SHIP1-deficient myeloid cells in the inflammatory disease that influence the lungs and gut. On the contrary, we believe that SHIP1-deficient myeloid cells ultimately cause the tissue destruction observed the *Ship1*[−/−] mice, but that their participation in this pathology is subsequent to, and a consequence of, a paucity of T effector function at these sites. These findings have important clinical implications, as alterations in SHIP1/INPP5D expression or function may contribute to human CD. Indeed our analysis of a small cohort of CD patients indicates a subset of these patients (10–15%) have little or no detectable SHIP1 protein expression and enzyme activity in their peripheral blood mononuclear cells (Fuhler, Fernandes, Peppelenbosch, Kerr, unpublished data). Another study is currently being organized to evaluate these initial findings in a larger patient cohort. The findings also suggest a selective role for SHIP1 in the persistence of effector T cells at mucosal sites; and thus one could consider using recently identified small molecule inhibitors of SHIP1[8,9] to target autoreactive T cells in other forms of IBD in which SHIP1 deficiency does not play a role, such as ulcerative colitis. Our preliminary studies indicate that this function of SHIP1 appears to involve protecting mucosal T cells from caspase 8-mediated cell death triggered by Fas–FasL interaction (Park, Srivastava, Sudan, Kerr, unpublished data) (Fig. 1C).

In sum, recent findings illustrate the diverse nature of SHIP1 interactions in various cellular and molecular contexts that are only beginning to be unraveled. We believe that careful and continued dissection of the different pathological mechanisms involved in these diseases will provide novel opportunities for therapeutic intervention targeting SHIP1.

Acknowledgments

This work was supported in part by grants from the NIH (RO1 HL72523, R01 HL085580, R01 HL107127) and the Paige Arnold Butterfly Run. W.G.K. is the Murphy Family Professor of Children's Oncology Research, an Empire Scholar of the State University of New York and a Senior Scholar of the Crohn's and Colitis Foundation of America.

Conflicts of interest

W.G. Kerr has patents pending and issued concerning the modulation and detection of SHIP activity in disease. The other authors have no conflicts to disclose.

References

1. Kerr, W.G. Inhibitor and activator: dual functions for SHIP in immunity and cancer. *Ann. N.Y. Acad. Sci.*
2. Hazen, A.L. *et al.* 2009. SHIP is required for a functional hematopoietic stem cell niche. *Blood* **113:** 2924–2933.
3. Ming-Lum, A. *et al.* 2012. A pleckstrin homology-related domain in SHIP1 mediates membrane localization during Fcgamma receptor-induced phagocytosis. *FASEB J.* **26:** 3163–3177.
4. Ong, C.J. *et al.* 2007. Small-molecule agonists of SHIP1 inhibit the phosphoinositide 3-kinase pathway in hematopoietic cells. *Blood* **110:** 1942–1949.
5. Zhang, J., S.F. Walk, K.S. Ravichandran & J.C. Garrison. 2009. Regulation of the Src homology 2 domain-containing inositol 5′-phosphatase (SHIP1) by the cyclic AMP-dependent protein kinase. *J. Biol. Chem.* **284:** 20070–20078.

6. Wahle, J.A. *et al.* 2007. Inappropriate recruitment and activity by the Src homology region 2 domain-containing phosphatase 1 (SHP1) is responsible for receptor dominance in the SHIP-deficient NK cell. *J. Immunol.* **179:** 8009–8015.

7. Peng, Q. *et al.* TREM2- and DAP12-dependent activation of PI3K requires DAP10 and is inhibited by SHIP1. *Sci. Signal.* **3:** ra38.

8. Brooks, R. *et al.* 2010. SHIP1 inhibition increases immunoregulatory capacity and triggers apoptosis of hematopoietic cancer cells. *J. Immunol.* **184:** 3582–3589.

9. Fuhler, G.M. *et al.* 2012. Therapeutic potential of SH2 domain-containing inositol-5′-phosphatase 1 (SHIP1) and SHIP2 inhibition in cancer. *Mol. Med.* **18:** 65–75.

10. Gewinner, C. *et al.* 2009. Evidence that inositol polyphosphate 4-phosphatase type II is a tumor suppressor that inhibits PI3K signaling. *Cancer Cell* **16:** 115–125.

11. Charlier, E. *et al.* 2010. SHIP-1 inhibits CD95/APO-1/Fas-induced apoptosis in primary T lymphocytes and T leukemic cells by promoting CD95 glycosylation independently of its phosphatase activity. *Leukemia* **24:** 821–832.

12. Helgason, C.D. *et al.* 1998. Targeted disruption of SHIP leads to hemopoietic perturbations, lung pathology, and a shortened life span. *Genes Develop.* **12:** 1610–1620.

13. Kerr, W.G., M.Y. Park, M. Maubert & R.W. Engelman. SHIP deficiency causes Crohn's disease-like ileitis. *Gut* **60:** 177–188.

14. Ghansah, T. *et al.* 2004. Expansion of myeloid suppressor cells in SHIP-deficient mice represses allogeneic T cell responses. *J. Immunol.* **173:** 7324–7330.

15. Liu, Q. *et al.* 1999. SHIP is a negative regulator of growth factor receptor-mediated PKB/Akt activation and myeloid cell survival. *Genes Develop.* **13:** 786–791.

16. Wang, J.W. *et al.* 2002. Influence of SHIP on the NK repertoire and allogeneic bone marrow transplantation. *Science* **295:** 2094–2097.

17. Karlsson, M.C. *et al.* 2003. Macrophages control the retention and trafficking of B lymphocytes in the splenic marginal zone. *J. Exp. Med.* **198:** 333–340.

Ann. N.Y. Acad. Sci. ISSN 0077-8923

ANNALS OF THE NEW YORK ACADEMY OF SCIENCES

Issue: *Inositol Phospholipid Signaling in Physiology and Disease*

Role of SHIP1 in bone biology

Sonia Iyer,[1] Bryan S. Margulies,[2] and William G. Kerr[1,2,3,4]

[1]Department of Microbiology and Immunology. [2]Department of Orthopedic Surgery. [3]Department of Pediatrics, SUNY Upstate Medical University, Syracuse, New York. [4]Department of Chemistry, Syracuse University, Syracuse, New York

Address for correspondence: William G. Kerr, Ph.D., SUNY Upstate Medical University, 750 E. Adams Street, 2204 Weiskotten Hall, Syracuse, NY 13210. kerrw@upstate.edu

The bone marrow milieu comprising both hematopoietic and nonhematopoietic lineages has a unique structural organization. Bone undergoes continuous remodeling throughout life. This dynamic process involves a balance between bone-forming osteoblasts (OBs) derived from multipotent mesenchymal stem cells (MSCs) and bone-resorbing osteoclasts (OCs) derived from hematopoietic stem cells (HSCs). Src homology 2-domain–containing inositol 5'-phosphatase 1 (SHIP1) regulates cellular processes such as proliferation, differentiation, and survival via the PI3K/Akt signaling pathway initiated at the plasma membrane. SHIP1-deficient mice also exhibit profound osteoporosis that has been proposed to result from hyperresorptive activity by OCs. We have previously observed that SHIP1 is expressed in primary OBs, which display defective development in SHIP1-deficient mice. These findings led us to question whether SHIP1 plays a functional role in osteolineage development from MSC *in vivo*, which contributes to the osteoporotic phenotype in germline SHIP1 knockout mice. In this short review, we discuss our current understanding of inositol phospholipid signaling downstream of SHIP1 in bone biology.

Keywords: SHIP1; bone; osteoclast; osteoblast; PI3K; inositol phospholipid

Introduction

The Src homology 2-domain-containing inositol 5'-phosphatase 1 (SHIP1) regulates cellular processes such as proliferation, differentiation, and survival via the PI3K/Akt signaling pathway initiated at the plasma membrane. SHIP1 dephosphorylates the product of PI3K, phosphatidylinositol-(3,4,5)-triphosphate ($PI(3,4,5)P_3$), to phosphatidylinositol-(3,4)-bisphosphate ($PI(3,4)P_2$), which, like $PI(3,4,5)P_3$, can facilitate downstream activation of Akt.[1] The SHIP1 protein encoded by the INPP5D gene was simultaneously identified as an LPS-response gene in B cells that binds both the SH3 domain of Grb2 via its polyproline motifs (PxxP) and the PTB domain of Shc via its NPXY motifs.[1] Furthermore, by virtue of its SH2-domain, this phosphatase can be recruited to receptor-associated signaling complexes at the plasma membrane directly or via either adaptor proteins, such as Shc, Grb2, Dok3, or by scaffolding proteins, such as Gab1/2. SHIP1 arrests DAP12-mediated activation of macrophages and osteoclasts by preventing the recruitment of other SH2 domain–containing proteins, including p85 and Syk, to the ITAM of DAP12.[2] Regulation of SHIP protein expression occurs both at the transcriptional and posttranscriptional levels, while SHIP1 protein can also be targeted for proteasomal degradation via ubiquitination.[1]

SHIP1 in osteoclasts

Bone undergoes continuous remodeling in the body throughout life. This dynamic process involves a balance between bone-forming osteoblasts (OBs), derived from multipotent mesenchymal stem cells (MSCs), and bone-resorbing osteoclasts (OCs), derived from hematopoietic stem cells (HSCs).[3] A balance of these immune and bone cells is required for maintaining normal bone homeostasis.[4] In addition to developing severe mucosal inflammatory disease in the lungs and terminal ileum,[5] *Ship*$^{-/-}$ mice exhibit profound osteoporosis that was proposed to result from hyper-resorptive behavior by OCs.[6] The overall resorptive activity of OCs, resulting from increased monocyte proliferation coupled with OC differentiation and survival, was attributed to increased PI3K/Akt activation downstream of TREM2,[2] increased D-type cyclins and downregulation of p27 in response to M-CSF,[7] and to the nonenzymatic functions of SHIP1 that lead to unopposed PI3K signaling at DAP12-associated receptors.[2]

SHIP1 in mesenchymal stem cells and osteoblasts

The bone marrow milieu, responsible for both hematopoietic and nonhematopoietic lineages, has a unique structural organization. Previously, we and others found that HSCs from SHIP-deficient mice demonstrate defective repopulating and self-renewal capacity upon transfer to SHIP-competent hosts.[8] These findings suggested that SHIP deficiency leads to an intrinsic defect in HSC function; however, this defect was not observed when HSCs were rendered SHIP-deficient in adult hosts where the BM milieu remained SHIP competent.[9] We therefore hypothesized that defective HSC function in germline SHIP-deficient mice might arise from disruption of niche cell components. Consistent with this hypothesis, primary osteoblasts were found to express the SH2 domain–containing 145 and 150 kD isoforms encoded by the SHIP1 locus.[9] In addition, bone marrow–derived *Ship1*$^{-/-}$ osteoblasts, grown *ex vivo* apart from *Ship*$^{-/-}$ OCs, expressed less alkaline phosphatase (ALP) activity, which is required for bone formation by OBs. This suggests that OB development and function might be directly impaired by SHIP deficiency.[9] However, the cellular components and molecular pathways that constitute the BM microenvironment altered by SHIP deficiency remained to be defined. Nonetheless, these findings strongly suggest that inositol phospholipid signaling pathways are critical to the function of the BM microenvironment that supports HSCs.

Mesenchymal stem cells are part of the BM niche that supports HSCs.[10] In addition, these multipotent stem cells can differentiate into various cell types that include HSC-supportive osteoblasts, as well as chondrocytes, adipocytes, or myocytes depending on signals derived from the cellular environment. Bone morphogenetic proteins (BMP) and transforming growth factor-β (TGF-β) play opposing roles in regulating MSC differentiation to OBs. For instance, BMP signaling stimulates osteoblast differentiation, while TGF-β suppresses

osteoblast differentiation and maturation.[11,12] Alliston *et al.* demonstrated that the transcriptional repression mechanism of core-binding factor α1 (Cbfa1), a key regulator of OB function, proliferation, and differentiation,[13] is mediated by SMAD3 and SMAD4 in a TGF-β–dependent manner (Fig. 1). Furthermore, canonical Wnt/β-catenin signaling regulates different stages of osteoblast differentiation, and evidence suggests that TGF-β regulates β-catenin signaling via PI3K during osteoblastogenesis of human MSC (Fig. 1).[14] Canonical Wnt/β-catenin signaling has been shown to be required for skeletal development;[15] however, some evidence suggests that in specific contexts, Wnt/β-catenin pathway activation can also suppress osteoblast differentiation and terminal differentiation.[16]

Our previous findings that SHIP1 expression occurs in primary OBs and that OB development is defective in SHIP1-deficient mice[9] led us to question whether SHIP1 plays a functional role during *in vivo* osteolineage development from MSCs that contributes to the osteoporotic phenotype of SHIP1-deficient mice.[6] We hypothesized that osteolineage expression of SHIP1 is required for efficient development of OBs, such that normal body growth, bone formation, and mineralization might be impaired in mice that lack osteolineage expression of SHIP1. Osteoblasts are key producers of factors such as M-CSF and RANKL that modulate osteoclastogenesis.[17] Therefore, either deletion of SHIP1 from OBs or MSC-producing OBs might have an effect on osteoclast differentiation. In addition, osteoblastic cells are important regulatory components of the HSC microenvironment.[18] It will thus be interesting to elucidate how SHIP1 in the osteoblastic niche is thought to maintain long-term HSCs and support their self-renewal.[18]

We speculate that SHIP1 may limit signaling pathways in MSCs that promote their proliferation and/or survival, and potentially facilitate MSC differentiation toward an osteoblast fate. Overexpression of the Id2 (inhibitor of differentiation 2) transcription factor has been shown to promote proliferation of MSCs while selectively blocking their osteolineage differentiation.[19] Moreover, coordinate overexpression of the deubiquitinase USP1 has been shown to be necessary for preventing proteasomal degradation and sustaining expression of

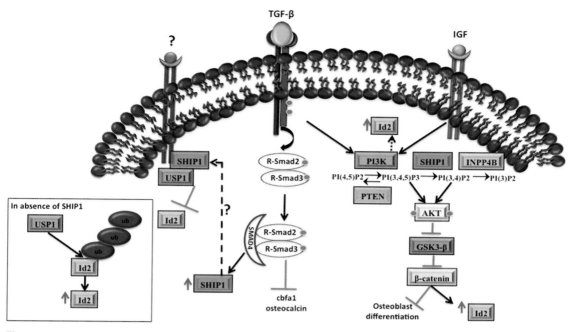

Figure 1. Role of SHIP1 in bone biology. TGF-β and IGF signaling can be mediated via the PI3K pathway, where AKT is regulated by both the SHIP1 substrate $PI(3,4,5)P_3$ and product $PI(3,4)P_2$. AKT signaling leads to GSK3β inhibition, which increases β-catenin levels and subsequently increases Id2 levels. The SMAD pathway is also activated by the TGF-β receptor complex, which represses cbfa1 and osteocalcin, and thus inhibits osteoblast differentiation. The SMAD complex can upregulate SHIP1, which binds to USP1 and blocks Id2. In the absence of SHIP1, USP1 deubiquitinates Id2 and prevents its degradation, leading to increased MSC proliferation and less differentiation to OB.

Id2 at levels sufficient to promote MSC proliferation and blockade of osteoblast differentiation.[19] In HSCs, upregulation of Id2 by β-catenin also suppresses myeloid differentiation and expands the hematopoietic precursor population.[20] We suspect that SHIP1 limits Id2 levels in MSCs, thus facilitating osteoblast development, potentially by sequestering USP1 and thereby preventing its interaction with Id2 (Fig. 1). In support of this hypothesis, we have observed significantly increased Id2 expression in SHIP1-deficient MSCs both before and upon osteogenic induction (Iyer, Margulies, and Kerr, unpublished observation).

Bone morphogenic proteins (BMPs) that promote either MSC cycling or osteolineage differentiation do so by activating SMAD family transcription factors, including SMAD4.[21] BMPs have also been shown to induce expression of Ids in a SMAD4-dependent fashion.[21] As induction of SMAD family transcription factors by TGF-β and LPS can induce SHIP1 expression,[22] we propose that SMAD4-mediated induction of SHIP1, downstream of multiple osteogenic factors, could serve as a molecular switch to promote osteolineage commitment by MSCs (Fig. 1). SMAD4 induction of SHIP1 in MSCs/osteoprogenitors would then provide a negative feedback loop for the effects of BMPs on MSCs by repressing USP1/Id2 and thus limiting MSC cycling, which promotes their differentiation to osteoblasts. Therefore, SHIP1 may regulate a switch that controls the USP1/Id2 axis, and thus influences stem versus lineage commitment of MSCs. If this model can be supported by additional experimental evidence, it would represent the first molecularly defined role for SHIP1 in the control of stem cell fate and proliferation.

Acknowledgments

This work was supported in part by grants from the NIH (RO1-HL72523, R01- HL085580, R01-HL107127) and the Paige Arnold Butterfly Run. W.G.K. is the Murphy Family Professor of Children's Oncology Research, an Empire Scholar of the State University of NY, and a Senior Scholar of the Crohn's and Colitis Foundation of America.

Conflicts of interest

W.G.K. has patents, both pending and issued, concerning the analysis and targeting of SHIP1 in disease. The other authors have no conflicts to disclose.

References

1. Kerr, W.G. 2011. Inhibitor and activator: dual functions for SHIP in immunity and cancer. *Ann. N.Y. Acad. Sci.* **1217:** 1–17.
2. Peng, Q. *et al.* 2010. TREM2- and DAP12-dependent activation of PI3K requires DAP10 and is inhibited by SHIP1. *Sci. Signal* **3:** ra38.
3. Hadjidakis, D.J. & I. Androulakis. 2006. Bone remodeling. *Ann. N.Y. Acad. Sci.* **1092:** 385–396.
4. Caetano-Lopes, J., H. Canhao & J.E. Fonseca. 2009. Osteoimmunology—the hidden immune regulation of bone. *Autoimmun. Rev.* **8:** 250–255.
5. Kerr, W.G., M.Y. Park, M. Maubert & R.W. Engelman. 2011. SHIP deficiency causes Crohn's disease-like ileitis. *Gut* **60:** 177–188.
6. Takeshita, S. *et al.* 2002. SHIP-deficient mice are severely osteoporotic due to increased numbers of hyper-resorptive osteoclasts. *Nat. Med.* **8:** 943–949.
7. Zhou, P. *et al.* 2006. SHIP1 negatively regulates proliferation of osteoclast precursors via Akt-dependent alterations in D-type cyclins and p27. *J. Immunol.* **177:** 8777–8784.
8. Desponts, C., J.M. Ninos & W.G. Kerr. 2006. s-SHIP associates with receptor complexes essential for pluripotent stem cell growth and survival. *Stem Cells Dev.* **15:** 641–646.
9. Hazen, A.L. *et al.* 2009. SHIP is required for a functional hematopoietic stem cell niche. *Blood* **113:** 2924–2933.
10. Mendez-Ferrer, S. *et al.* 2010. Mesenchymal and haematopoietic stem cells form a unique bone marrow niche. *Nature* **466:** 829–834.
11. Maeda, S., M. Hayashi, S. Komiya, *et al.* 2004. Endogenous TGF-beta signaling suppresses maturation of osteoblastic mesenchymal cells. *EMBO J.* **23:** 552–563.
12. Alliston, T., L. Choy, P. Ducy, *et al.* 2001. TGF-beta-induced repression of CBFA1 by Smad3 decreases cbfa1 and osteocalcin expression and inhibits osteoblast differentiation. *EMBO J.* **20:** 2254–2272.
13. Karsenty, G. *et al.* 1999. Cbfa1 as a regulator of osteoblast differentiation and function. *Bone* **25:** 107–108.
14. Zhou, S. 2011. TGF-beta regulates beta-catenin signaling and osteoblast differentiation in human mesenchymal stem cells. *J. Cell Biochem.* **112:** 1651–1660.
15. Krishnan, V., H.U. Bryant & O.A. Macdougald. 2006. Regulation of bone mass by Wnt signaling. *J. Clin. Invest.* **116:** 1202–1209.
16. Liu, F., S. Kohlmeier & C.Y. Wang. 2008. Wnt signaling and skeletal development. *Cell Signal* **20:** 999–1009.
17. Tsurukai, T., N. Udagawa, K. Matsuzaki, *et al.* 2000. Roles of macrophage-colony stimulating factor and osteoclast differentiation factor in osteoclastogenesis. *J. Bone Miner. Metab.* **18:** 177–184.
18. Calvi, L.M. *et al.* 2003. Osteoblastic cells regulate the haematopoietic stem cell niche. *Nature* **425:** 841–846.
19. Williams, S.A. *et al.* 2011. USP1 deubiquitinates ID proteins to preserve a mesenchymal stem cell program in osteosarcoma. *Cell* **146:** 918–930.
20. Perry, J.M. *et al.* 2011. Cooperation between both Wnt/{beta}-catenin and PTEN/PI3K/Akt signaling promotes primitive hematopoietic stem cell self-renewal and expansion. *Genes Dev.* **25:** 1928–1942.
21. Peng, Y. *et al.* 2004. Inhibitor of DNA binding/differentiation helix-loop-helix proteins mediate bone morphogenetic protein-induced osteoblast differentiation of mesenchymal stem cells. *J. Biol. Chem.* **279:** 32941–32949.
22. Pan, H. *et al.* 2010. SMAD4 is required for development of maximal endotoxin tolerance. *J. Immunol.* **184:** 5502–5509.

Ann. N.Y. Acad. Sci. ISSN 0077-8923

ANNALS OF THE NEW YORK ACADEMY OF SCIENCES
Issue: *Inositol Phospholipid Signaling in Physiology and Disease*

Achieving cancer cell death with PI3K/mTOR-targeted therapies

Sung Su Yea[1] and David A. Fruman[2]

[1]Department of Biochemistry, College of Medicine, Inje University, Busan 614-735, Korea. [2]Department of Molecular Biology & Biochemistry, University of California, Irvine, Irvine, California

Address for correspondence: David A. Fruman, University of California, Irvine, Department of Molecular Biology & Biochemistry, 3242 McGaugh Hall, Irvine, CA 92697–3900. dfruman@uci.edu

Inhibitors of the PI3K/mTOR signaling network are under development as novel cancer therapies. However, these compounds do not cause robust cytotoxic responses in tumor cells unless combined with other agents. Rational combinations with other targeted therapies will likely be necessary to achieve the potential of PI3K/mTOR inhibitors in oncology.

Keywords: PI3K; mTOR; apoptosis; chemoresistance

Introduction

Phosphoinositide 3-kinase (PI3K) and the mechanistic target of rapamycin (mTOR) participate in a signaling network that drives cell proliferation and survival in response to growth factors or oncogenic changes.[1,2] Development of small molecule inhibitors of PI3K and/or mTOR has been an active area of drug discovery for cancer, and several compounds are in clinical trials. However, PI3K/mTOR inhibitors show limited ability to cause cancer cell death when used as single agents. In preclinical models, PI3K/mTOR inhibitors achieve tumor regression most effectively when used in combination with other targeted therapies.[3–5] Here we discuss models to explain cancer cell resistance to apoptosis, and speculate on rational combinations to unleash the proapoptotic potential of PI3K/mTOR inhibitors.

PI3K/mTOR signaling and cell survival

PI3Ks are a family of broadly expressed enzymes that produce 3-phosphorylated inositol lipids to promote membrane recruitment of specific effectors.[6] Among the different PI3K catalytic isoforms, the class I PI3Ks (PI3Kα, PI3Kβ, PI3Kγ, PI3Kδ) are mainly responsible for signals leading to cell growth, proliferation, and survival. Activating mutations in the gene encoding PI3Kα (*PIK3CA*) are frequent in human cancer and the other three class I isoforms are thought to promote cancer in certain contexts.[1] A key downstream effector of PI3K signaling is mTOR, a serine/threonine kinase encoded by a single gene, *MTOR*. The mTOR kinase is the catalytic subunit of two multiprotein complexes, TORC1 and TORC2, with different substrates and functions.[2] TORC1 and TORC2 together control a large fraction of the cellular phosphorylation responses to growth factors and mTOR substrate phosphorylation is elevated in most cancer cells.

Class I PI3K and TORC2 provide essential inputs to activate AKT (also known as PKB), a multifunctional serine/threonine kinase with many cellular substrates.[2,6] A large body of literature supports the idea that AKT promotes cell survival through phosphorylation of many proteins, including BAD, MDM2, and FOXO transcription factors.[7] However, much of the evidence for PI3K/AKT-mediated survival signaling came from early studies using nonspecific inhibitors such as wortmannin and LY294002. Development of compounds with selectivity toward class I PI3Ks and/or mTOR, or AKT, has prompted reconsideration of earlier models. Namely, most cell types (cancer or normal) do not display a dramatic cell death response when treated with inhibitors of PI3K, AKT or mTOR as

doi: 10.1111/nyas.12028

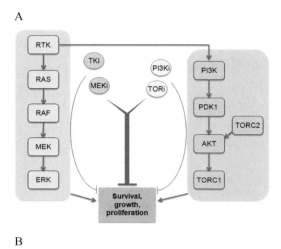

A

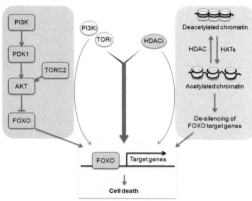

B

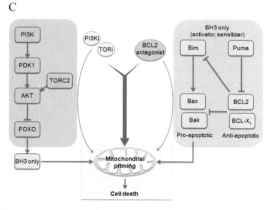

C

Figure 1. Pharmacological strategies to overcome resistance to apoptosis in cancer cells treated with PI3K/mTOR inhibitors. (A) Many tumor cells have activated receptor tyrosine kinases (RTKs), or cytoplasmic tyrosine kinases, and/or activation of the RAS/RAF/MEK/ERK pathway. Combining TKIs or MEK inhibitors with PI3K/mTOR inhibitors can induce cancer cell death. (B) Proapoptotic genes are often silenced epigenetically in tumor cells. HDAC inhibition leads to increased histone acetylation and a more open chromatin state at gene regulatory

single agents. In our experience, ATP-competitive mTOR inhibitors or dual PI3K/mTOR inhibitors can induce significant apoptosis in murine bone marrow cells transformed with the human oncogene BCR-ABL.[5] However, in bona fide human leukemia cells these compounds are more effective when used in combination with inhibitors of the BCR-ABL tyrosine kinase.[5,8] For scientists working in the PI3K/mTOR field, early optimism about single agent efficacy is evolving into the realization that effective application of PI3K/mTOR inhibitors in oncology will usually require rational combinations. Even in the case of CAL-101, a PI3Kδ inhibitor showing promising results in B cell malignancies, clinical trials suggest greater efficacy when CAL-101 is combined with rituximab or bendamustine.[1] How can we apply existing knowledge to design the best rational combinations?

Resistance model #1: compensatory signaling pathways

In most cancer cells, elevated PI3K/mTOR activity is not the only signaling mechanism that feeds into prosurvival mechanisms. The RAS/RAF/MEK/ERK pathway is activated in most cancer cells and acts in parallel to promote cell survival, sometimes converging on shared targets with PI3K/mTOR (Fig. 1A). Combinations of MEK inhibitors with PI3K/mTOR inhibitors have proved to be more effective than single agents in mouse models of *KRAS*-driven lung cancer[3] and other models. In many cases, the PI3K and RAS pathways are downstream of activated receptor tyrosine kinases (e.g., EGFR) or nonreceptor tyrosine kinases (e.g., BCR-ABL). Combining mTOR kinase inhibitors with tyrosine kinase inhibitors (TKIs) has shown promise in BCR-ABL–dependent leukemias, as mentioned above,[5,8] and in EGFR-dependent breast cancer.[4] Therefore, a general strategy for tumors with known lesions in tyrosine kinases might be to combine PI3K/mTOR inhibitors with TKIs or with MEK inhibitors (Fig. 1A).

elements. PI3K/mTOR inhibition triggers nuclear accumulation of FOXO transcription factors, which can access proapoptotic gene promoters when cells are treated with HDAC inhibitors. (C) Low mitochondrial priming is a significant barrier to apoptotic induction in some cancer cells. BCL2 antagonists can increase priming and lower the threshold for apoptotic induction by PI3K/mTOR inhibitors.

Resistance model #2: epigenetic status

It is becoming accepted that epigenetic changes play a major role in maintaining the transformed state of cancer cells and providing resistance to cancer therapies.[9] A simple model is that most apoptotic stimuli act by induction of prodeath genes, and such genes may be silenced epigenetically in cancer cells. This could occur by promoter methylation or through various posttranslational modifications of histones. Reversing the suppressive epigenetic marks might potentiate the apoptotic program.

One of the mechanisms for gene silencing is histone deacetylation (Fig. 1B). The concept of treating cancer with histone deacetylase inhibitors (HDACi) has gained acceptance with the approval of vorinostat for the treatment of cutaneous T cell lymphoma. A number of studies have reported synergistic killing of tumor cells with HDACi combined with inhibitors of the PI3K/mTOR network. Our laboratory has observed synergy of vorinostat with mTOR kinase inhibitors in B cell acute lymphoblastic leukemia (B-ALL) cells (D.A. Fruman, unpublished data). A possible mechanism is the desilencing of genes induced by FOXO transcription factors, which enter the nucleus when TORC2 and AKT are suppressed. Hence, we propose that HDACi should be explored more broadly for their ability to potentiate FOXO-dependent death when the PI3K/mTOR network is suppressed (Fig. 1B). Other drugs targeting the epigenome are also worth investigating if a molecular rationale exists.

Resistance model #3: mitochondrial priming

Apoptosis is an orderly process whose rate-limiting step is loss of mitochondrial membrane potential. Ultimately, the ability of drugs to induce cell death is determined by the balance of prosurvival and proapoptotic factors at the mitochondrial membrane. The closeness of cells to the threshold for apoptosis has been termed mitochondrial priming.[10] Recent studies have shown that the mitochondrial priming status is a major determinant of response to chemotherapy in a number of different malignancies. It is also likely that priming controls the response to targeted therapies. Consistent with this, CAL-101 treatment was shown to increase priming of CLL cells, potentiating apopto-

sis especially in the presence of chemotherapeutic agents.[10]

The main determinant of mitochondrial priming is the BCL2 family of apoptotic regulators (Fig. 1C). Elevated priming can be achieved by increased function of BH3-only family members, or by decreased function of prosurvival members BCL2, BCL-X$_L$, MCL-1, and others. Based on this knowledge, use of BCL2 antagonists (e.g., ABT-263) is a rational strategy for increasing the proapoptotic effect of various cancer drugs, including PI3K/mTOR inhibitors (Fig. 1C). For example, PI3K/mTOR inhibitors can increase expression of FOXO targets such as Bim and Puma, whose proapoptotic effects would be potentiated by increased priming through BCL2 antagonists.

Summary

Great progress has been achieved in PI3K/mTOR drug discovery. Small molecules with high selectivity and good drug properties have entered oncology trials after demonstrating proof of principle in animal models of cancer. However, optimism about this drug class has been tempered by the evidence that single agent PI3K/mTOR inhibitors are generally more cytostatic than cytotoxic. For these compounds to become a broadly used tool to eradicate cancer, it is essential to unleash the proapoptotic potential of inhibiting this pathway. A common strategy is to combine PI3K/mTOR inhibitors with compounds targeting other oncogenic signaling kinases, such as tyrosine kinases or MEK. Two other strategies discussed above are to combine PI3K/mTOR inhibitors with agents that either alter the cancer epigenome or increase mitochondrial priming. Many other strategies exist and deserve consideration.

On a cautionary note, it is likely that combinations that augment killing of cancer cells will also have greater toxicity to normal tissues. Identifying drug combinations with an acceptable therapeutic window will be a continuing challenge; but it is hoped that such combinations will eventually improve healthy survival of patients afflicted with cancer.

Acknowledgments

S.S.Y. was supported by the Inje Research and Scholarship Foundation in 2011. D.A.F. was supported by NIH R01-CA158383.

Conflicts of interest

The authors declare no conflicts of interest.

References

1. Fruman, D.A. & C. Rommel. 2011. PI3Kdelta Inhibitors in cancer: rationale and serendipity merge in the clinic. *Cancer Discov.* **1:** 562–572.
2. Laplante, M. & D.M. Sabatini. 2012. mTOR Signaling in Growth Control and Disease. *Cell* **149:** 274–293.
3. Engelman, J.A. *et al.* 2008. Effective use of PI3K and MEK inhibitors to treat mutant Kras G12D and PIK3CA H1047R murine lung cancers. *Nat. Med.* **14:** 1351–1356.
4. Garcia-Garcia, C. *et al.* 2012. Dual mTORC1/2 and HER2 blockade results in antitumor activity in preclinical models of breast cancer resistant to anti-HER2 therapy. *Clin. Cancer Res.* **18:** 2603–2612.
5. Janes, M.R. *et al.* 2010. Effective and selective targeting of leukemia cells using a TORC1/2 kinase inhibitor. *Nat. Med.* **16:** 205–213.
6. Vanhaesebroeck, B. *et al.* 2010. The emerging mechanisms of isoform-specific PI3K signalling. *Nat. Rev. Mol. Cell Biol.* **11:** 329–341.
7. Manning, B.D. & L.C. Cantley. 2007. AKT/PKB signaling: navigating downstream. *Cell* **129:** 1261–1274.
8. Janes, M.R. *et al.* 2012. Efficacy of the investigational mTOR kinase inhibitor MLN0128/INK128 in models of B-cell acute lymphoblastic leukemia. *Leukemia.* doi: 10.1038/leu.2012.276. Oct. 1 [Epub ahead of print].
9. Popovic, R. & J.D. Licht. 2012. Emerging epigenetic targets and therapies in cancer medicine. *Cancer Discov.* **2:** 405–413.
10. Davids, M.S. *et al.* 2012. Decreased mitochondrial apoptotic priming underlies stroma-mediated treatment resistance in chronic lymphocytic leukemia. *Blood* **120:** 3501–3509.

Ann. N.Y. Acad. Sci. ISSN 0077-8923

ANNALS OF THE NEW YORK ACADEMY OF SCIENCES
Issue: *Inositol Phospholipid Signaling in Physiology and Disease*

Challenges in the clinical development of PI3K inhibitors

Cristian Massacesi,[1] Emmanuelle di Tomaso,[2] Nathalie Fretault,[1] and Samit Hirawat[3]

[1]Novartis Oncology, Paris, France. [2]Novartis Institutes for BioMedical Research, Inc., Cambridge, Massachusetts. [3]Novartis Pharmaceuticals Corporation, Florham Park, New Jersey

Address for correspondence: Cristian Massacesi, Novartis Oncology, 14 Blvd Richelieu, 92500 Rueil-Malmaison, Paris, France. cristian.massacesi@novartis.com

The PI3K/Akt/mTOR pathway is one of the most frequently dysregulated signaling pathways in cancer and an important target for drug development. PI3K signaling plays a fundamental role in tumorigenesis, governing cell proliferation, survival, motility, and angiogenesis. Activation of the pathway is frequently observed in a variety of tumor types and can occur through several mechanisms. These mechanisms include (but are not limited to) upregulated signaling via the aberrant activation of receptors upstream of PI3K, amplification or gain-of-function mutations in the *PIK3CA* gene encoding the p110α catalytic subunit of PI3K, and inactivation of *PTEN* through mutation, deletion, or epigenetic silencing. PI3K pathway activation may occur as part of primary tumorigenesis, or as an adaptive response (via molecular alterations or increased phosphorylation of pathway components) that may lead to resistance to anticancer therapies. A range of PI3K inhibitors are being investigated for the treatment of different types of cancer; broad clinical development plans require a flexible yet well-structured approach to clinical trial design.

Keywords: PI3K antagonists; PI3K inhibitors; patient selection; research design

Introduction

The widespread and pivotal role of the PI3K pathway in cancer has inspired the active development of a spectrum of drugs that target various components of the pathway. These drugs include allosteric mTORC1 inhibitors, Akt inhibitors, inhibitors of all four class I PI3K isoforms (so-called pan-class I PI3K inhibitors), dual pan-class I PI3K and mTORC1/2 inhibitors, and, most recently, isoform-specific PI3K inhibitors. Novel compounds in clinical development by Novartis include the pan-PI3K inhibitor buparlisib (BKM120), the dual pan-PI3K/mTORC1/2 inhibitor BEZ235, and the selective p110α inhibitor BYL719. In addition, the mTORC1 inhibitor everolimus is already approved for use in several types of cancer (Fig. 1).

Due to the complexity of the PI3K pathway, and the extensive cross-talk with other pathways, one of the greatest challenges in PI3K inhibitor devel-

opment involves identifying the patients that will benefit most from treatment. Early-phase single-agent trials with PI3K inhibitors have yet to identify a consistent and distinct association between typical PI3K pathway alterations (*PIK3CA* mutation and PTEN loss) and response to therapy. This may partly be due to the heterogeneous range of cancers treated in these trials. The PI3K pathway interacts with other signaling pathways at several points, and these interactions are known to vary in a tissue-specific manner. Therefore, the capability of predictive biomarkers, and the effectiveness of different types of PI3K inhibitors, may also vary across tumor types. As the development of PI3K inhibitors progresses from mid to late phase and expands into tumor-specific studies, Novartis is employing a flexible approach to biomarker-driven study design, using a range of strategies based on the phase of drug development, the type of PI3K inhibitor, the tumor type under investigation, and the specific context of treatment. This mini-review

doi: 10.1111/nyas.12060

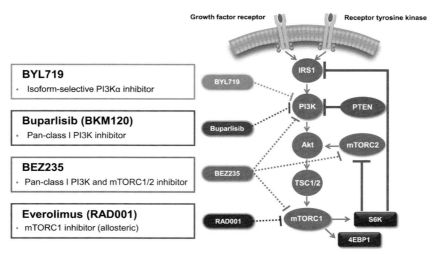

Figure 1. The PI3K/Akt/mTOR pathway and inhibitors that target it. IRS1, insulin receptor substrate 1; mTORC, mammalian target of rapamycin complex; PI3K, phosphatidylinositol 3-kinase; PTEN, phosphatase and tensin homolog; TSC, tuberous sclerosis protein.

summarizes four distinct approaches to study design and describes the rationale for their use in terms of the currently enrolling trials with Novartis PI3K inhibitors.

Patient stratification based on PI3K pathway status (breast cancer)

PI3K inhibitors have demonstrated encouraging preliminary activity in the treatment of metastatic breast cancer, with responses observed in patients with and without *PIK3CA* and *PTEN* alterations.[1,2] Evidence for the activity of PI3K inhibitor–based therapy in breast cancer has been drawn from a phase I study in patients with hormone receptor (HR)–positive metastatic breast cancer.[3] In this trial, patients received continuous ($n = 20$) or intermittent (five days on, two days off; $n = 31$) doses of buparlisib in combination with letrozole. The majority of patients ($n = 43$) had received prior aromatase-inhibitor therapy. The clinical benefit rate (complete responses plus partial responses plus stable disease) at six months was 30% and 29% in the continuous and intermittent cohorts, respectively. A correlation between duration of response or clinical benefit and the presence of *PIK3CA* mutation has yet to be observed in either cohort.

Given the aforementioned findings, the approach Novartis has taken in breast cancer has been to develop trials that are adequately powered to prospectively investigate efficacy in both the population as a whole and in the subpopulation of patients with PI3K pathway alterations. BELLE-2 (NCT01610284) is a multicenter phase III, placebo-controlled study of buparlisib plus fulvestrant that will enroll 842 postmenopausal women with HR-positive/HER2-negative advanced breast cancer whose disease has progressed on or after aromatase-inhibitor therapy, including $\geq$ 334 patients with PI3K pathway alterations. Enrollment will be stratified by the presence or absence of PI3K pathway activation, defined as *PIK3CA* mutation and/or *PTEN* alteration. BELLE-2 is designed to investigate progression-free survival (PFS) in the population as a whole and/or in the PI3K pathway-activated subpopulation using a gate-keeping procedure based on a graphical approach to address the multiplicity of hypotheses.[4] The results of this study could provide prospective evidence regarding the use of these biomarkers in predicting response to PI3K inhibitor therapy. Other trials with buparlisib in breast cancer are employing similar approaches, including a placebo-controlled phase II trial with paclitaxel in the first-line treatment of HER2-negative metastatic breast cancer (BELLE-4; NCT01572727), and a phase II trial of neoadjuvant paclitaxel plus trastuzumab, with and without buparlisib (Neo-PHOEBE) in HER2-overexpressing breast cancer patients.

Nonselective enrollment and mandatory tissue collection (prostate cancer and glioblastoma)

Another strategy is to conduct early-phase trials in tumor types with high frequencies of PI3K pathway alterations and strong preclinical evidence supporting the potential efficacy of PI3K-inhibition treatment. These trials enroll patients regardless of PI3K pathway status; however, enrollment is dependent upon the mandatory provision of tumor tissue, which can be used for exploratory *post hoc* analyses. Castration-resistant prostate cancer (CRPC) is one such tumor type being investigated using this strategy.

PTEN loss is one of the most frequent molecular aberrations to occur in prostate cancer, and ~70% of metastatic cases have some form of alteration in the PI3K pathway. This high frequency of alterations supports the rationale for investigating PI3K inhibitors in this tumor type. Furthermore, interaction and reciprocal feedback regulation between the androgen receptor and PI3K pathways has been suggested as a potential mechanism of resistance to androgen-deprivation therapy in CRPC. PI3K inhibitors may therefore have the potential to reverse resistance in this context. In preclinical experiments, the combination of BEZ235 and enzalutamide (an androgen-receptor antagonist) demonstrated near-complete tumor regression in a PTEN-deficient murine model and in human prostate cancer xenografts.[5] A phase Ib proof-of-concept trial of BEZ235 or buparlisib in combination with abiraterone acetate is currently enrolling patients with CRPC after progression on abiraterone acetate (NCT01634061).

Glioblastoma multiforme (GBM) is another tumor type with a high frequency of PI3K pathway alterations, with PTEN loss reported in up to 35% of cases. Buparlisib has demonstrated an ability to cross the blood–brain barrier and inhibit the PI3K pathway in the brain, and has shown synergy with temozolomide and docetaxel in murine xenografts of *PTEN*-null GBM.[6] A phase I trial is investigating buparlisib in combination with adjuvant temozolomide and with concomitant radiotherapy and temozolomide in newly diagnosed GBM (NCT01473901). Two other ongoing phase

I/II trials are investigating single-agent buparlisib or the combination of buparlisib and bevacizumab in patients with relapsed disease (NCT01339052 and NCT01349660, respectively). Enrollment in both of these trials is dependent on the provision of tumor biopsy material for the analysis of PI3K pathway alterations.

Preselection of patients with PI3K pathway activation—enrichment strategy (nonsmall cell lung cancer)

Certain contexts may necessitate the design of trials that selectively recruit patients with PI3K pathway alterations only. Lung cancer treatment has recently moved toward a customized approach based on the molecular characteristics of tumors: patients with *EGFR* mutations may show improved benefit from EGFR tyrosine kinase inhibitors (TKIs; e.g., erlotinib and gefitinib), and those with *ALK* translocations from ALK inhibitors (e.g., crizotinib). Preclinical experiments have suggested that PI3K pathway alterations may predict a differential response to PI3K inhibitors in models of nonsmall cell lung cancer (NSCLC),[7] and PI3K pathway activation has been identified as one of the factors driving resistance to EGFR TKIs in preclinical models.[8] An ongoing phase II study (NCT01297491) is therefore evaluating single-agent buparlisib versus docetaxel or pemetrexed in patients with squamous or nonsquamous metastatic NSCLC with PI3K pathway alterations (*PIK3CA* mutation and/or *PTEN* alteration). Patients who have been pretreated with one or two prior antineoplastic treatments are eligible.

Isoform-specific PI3K inhibitors may theoretically offer an improved therapeutic window and narrower toxicity profile compared with pan-PI3K inhibitors. The selective PI3Kα inhibitor BYL719 has shown preferential sensitivity in *PIK3CA*-mutated cell lines, and a first-in-man study with this agent (NCT01387321) is enrolling patients with *PIK3CA* mutation or amplification only to maximize the potential benefit of treatment.[9] Preliminary results from this phase I trial of single-agent BYL719 in patients with advanced solid tumors suggests a favorable safety profile, with two confirmed partial responses observed (one each in patients with HR-positive breast cancer and cervical cancer).[9]

Enrollment of patients that have progressed on mTORC1 inhibitor-based therapy

The BOLERO-2 trial showed substantial improvements in PFS with the combination of everolimus and exemestane, compared with exemestane alone, in patients with advanced HR-positive breast cancer who had progressed on nonsteroidal aromatase inhibitors.[10] Despite these improvements in PFS, resistance to the combination of everolimus and exemestane can occur. Inhibition of mTORC1, but not mTORC2, can cause paradoxical reactivation of the PI3K pathway through the alleviation of feedback loops dependent on S6K.[11] PI3K inhibitors, which target the pathway upstream of mTORC1, may therefore show utility in contexts in which mTORC1 inhibitors are unsuccessful or no longer effective. The potential use of PI3K inhibitors in the post-mTORC1 inhibitor treatment setting is being investigated in BELLE-3 (NCT01633060), a placebo-controlled phase III study to investigate the safety and efficacy of buparlisib plus fulvestrant in postmenopausal women with HR-positive/HER2-negative advanced breast cancer who have received aromatase-inhibitor treatment and progressed on or after mTORC1 inhibitor-based therapy. Like BELLE-2, BELLE-3 is stratifying enrolling patients according to PI3K pathway activation status, to investigate the treatment effect in patients with PI3K pathway activation and/or the population as a whole.

Summary

The burgeoning field of PI3K inhibitor development is associated with many ongoing challenges. PI3K signaling is complex and can be modulated by crosstalk with other kinase cascades, such as the Ras/Raf/MEK pathway. This complexity is further compounded by tissue-specific effects, which may complicate the identification of predictive biomarkers.

It remains unclear whether preclinical observations of improved responses to PI3K inhibitors in tumors with *PIK3CA* and *PTEN* alterations will be borne out in clinical trials. Early-phase, single-agent trials with PI3K inhibitors have yet to establish a consistent and distinct association between the most common alterations in the PI3K pathway and response to therapy. Explanations for this are numerous and include heterogeneity in the patient population, use of archival specimens for biomarker assessment, and a low number of responses to single-agent PI3K inhibitors. Future trials of PI3K inhibitors as combination therapy in more homogenous patient populations may be more likely to establish a link between typical PI3K alterations and clinical response.

The successful evaluation of PI3K pathway biomarkers is complicated by many factors, such as observations of discordance between primary and metastatic lesions and issues with intratumoral heterogeneity in molecular alterations. It is possible that future studies will require the prospective collection of biopsies immediately before and after study treatment to address these difficulties. Advances in noninvasive technologies, such as circulating DNA and/or tumor cell analysis, may eventually allow this approach. Future studies may also benefit from deeper analyses into pathway alterations and signaling, such as those offered by high-throughput, next-generation sequencing and phosphoproteomic analyses.

Finally, the spectrum of different PI3K inhibitors also presents its own dilemma: How are the optimal indications for each class of inhibitor identified? Whereas more selective inhibitors may offer improved therapeutic windows and narrower toxicity profiles, certain tumor types or treatment contexts may require more comprehensive inhibition of the PI3K pathway. These assessments will include the identification of the optimal dose and dosing schedule of each inhibitor and the tumor types in which they can best be used, and will be planned based on robust preclinical evidence.

In conclusion, PI3K inhibitors show great promise in the treatment of a wide range of cancers; a well-structured approach to study design will be required to maximize the potential of this exciting class of therapy.

Acknowledgments

Financial support for medical editorial assistance was provided by Novartis Pharmaceuticals. Ben Holtom, D.Phil., is thanked for his medical editorial assistance with this manuscript.

Conflicts of interest

Cristian Massacesi, Emmanuelle di Tomaso, Nathalie Fretault, and Samit Hirawat are employees of Novartis Pharmaceuticals Corp.

References

1. Bendell, J.C., J. Rodon, H.A. Burris, *et al.* 2012. Phase I, dose-escalation study of BKM120, an oral pan-class I PI3K inhibitor, in patients with advanced solid tumors. *J. Clin. Oncol.* **30:** 282–290.

2. Krop, I., C. Saura, J. Rodon, *et al.* 2012. A Phase I/IB dose-escalation study of BEZ235 in combination with trastuzumab in patients with PI3-kinase or PTEN altered HER2$^+$ metastatic breast cancer. *J. Clin. Oncol.* **30:** 508 (abstract).

3. Mayer, I. A., V.G. Abramson, J.M. Balko, *et al.* 2012. SU2C phase Ib study of pan-PI3K inhibitor BKM120 with letrozole in ER$^+$/HER2- metastatic breast cancer (MBC). *J. Clin. Oncol.* **30:** 510 (abstract).

4. Bretz, F., M. Posch, E. Glimm, *et al.* 2011. Graphical approaches for multiple comparison procedures using weighted Bonferroni, Simes, or parametric tests. *Biom. J.* **53:** 894–913.

5. Carver, B.S., C. Chapinski, J. Wongvipat, *et al.* 2011. Reciprocal feedback regulation of PI3K and androgen receptor signaling in PTEN-deficient prostate cancer. *Cancer Cell.* **19:** 575–586.

6. Maira, S.M., S. Pecchi, A. Huang, *et al.* 2012. Identification and characterization of NVP-BKM120, an orally available pan-class I PI3-Kinase inhibitor. *Mol. Cancer Ther.* **11:** 317–328.

7. Engelman, J.A., L. Chen, X. Tan, *et al.* 2008. Effective use of PI3K and MEK inhibitors to treat mutant Kras G12D and PIK3CA H1047R murine lung cancers. *Nat. Med.* **14:** 1351–1356.

8. Sos, M.L., M. Koker, B.A. Weir, *et al.* 2009. PTEN loss contributes to erlotinib resistance in EGFR-mutant lung cancer by activation of Akt and EGFR. *Cancer Res.* **69:** 3256–3261.

9. Juric, D., J. Rodon, A.M. Gonzalez-Angulo, *et al.* 2012. BYL719, a next generation PI3K alpha specific inhibitor: preliminary safety, PK, and efficacy results from the first-in-human study. In *Proceedings of the 103rd Annual Meeting of the American Association for Cancer Research (March 31–April 4)*. Chicago, IL. AACR. Philadelphia (PA). Abstract nr CT-01.

10. Baselga, J., M. Campone, M. Piccart, *et al.* 2012. Everolimus in postmenopausal hormone-receptor-positive advanced breast cancer. *New Engl. J. Med.* **366:** 520–529.

11. Markman, B., R. Dienstmann & J. Tabernero. 2010. Targeting the PI3K/Akt/mTOR pathway—beyond rapalogs. *Oncotarget.* **1w:** 530–543.

Ann. N.Y. Acad. Sci. ISSN 0077-8923

ANNALS OF THE NEW YORK ACADEMY OF SCIENCES

Issue: *Inositol Phospholipid Signaling in Physiology and Disease*

Rules of engagement: distinct functions for the four class I PI3K catalytic isoforms in immunity

Klaus Okkenhaug

Laboratory of Lymphocyte Signaling and Development, The Babraham Institute, Cambridge, United Kingdom

Address for correspondence: Klaus Okkenhaug, Laboratory of Lymphocyte Signaling and Development, The Babraham Institute, Cambridge, CB22 3AT, UK. Klaus.okkenhaug@babraham.ac.uk

Mammalian cells can express up to four different class I phosphatidylinositide 3-kinase (PI3K) isoforms, each of which is engaged by tyrosine kinases or G protein–coupled receptors (GPCRs) to generate the second messenger signaling molecule PtdIns(3,4,5)P$_3$ (PIP$_3$). The p110α and p110β isoforms are relatively widely expressed, whereas p110γ and p110δ are more highly expressed in cells of the immune system than in other cell types. Nevertheless, each of the four class I PI3Ks have been shown to participate in the orchestration of the signaling events that lead to immune cell development and control of gene expression, skewing toward individual cell lineage subsets and proliferation.

Keywords: PI3K; PIP$_3$; T cell; B cell; PI3K isoforms

Introduction

Because they are constitutively bound by p85-related regulatory subunits, the phosphatidylinositide 3-kinase (PI3K) isoforms p110α, p110β and p110δ were thought to be regulated primarily by tyrosine kinases (p85 contains two SH2 domains that bind tyrosine phosphorylated peptides), whereas the p110γ isoform, which is bound by either a G$_{\beta\gamma}$-binding p101 or p84 regulatory subunit, is preferentially activated by G protein–coupled receptors (GPCRs). However, by mechanisms that have yet to be characterized, p110β and p110δ can also be activated by GPCRs. Over the last few years, experiments combining the use of gene-targeted mice and small molecule inhibitors have identified individual roles for each of the four class I PI3K catalytic subunits in immunity.

Role of p110α and p110δ in B cell development and survival

During early B cell development, p110δ plays a redundant role with p110α to promote further differentiation after the pre-B cells have rearranged the immunoglobulin heavy chain that forms part of the pre-B cell receptor complex.[1] Remarkably, the expression of a p110α from a single allele is sufficient to allow B cell development beyond this checkpoint in the bone marrow in absence of p110δ activity. By contrast, in mature B cells, p110α does not contribute significantly to signaling by the mature B cell receptor that is composed of immunoglobulin heavy and light chains. The molecular basis of the reduced role of p110α as B cells mature is not thought to be due to reduced expression of this isoform, rather, it appears that p110α is less well adapted to respond to acute signaling induced by clustering of the BCR rather than by so-called tonic signaling. Accordingly, B1 and marginal zone B cells, which are thought be autoreactive, develop in the absence of p110α but not in the absence of p110δ activity.[1,2] However, pre-B cell development and mature B cell survival, which are driven by tonic signaling by BCR in absence of agonistic antigen, are dependent on both p110α and p110δ.

Role of p110δ and p110γ in T cell development

T cells retain the ability to develop in the thymus and populate the spleen and lymph nodes, despite loss of p110α and p110δ.[1] However, when p110δ and p110γ were both lost, T cell development

doi: 10.1111/nyas.12027

SOCS3 expression, and inhibits Th1 and Th17 differentiation.[8] Although deletion of *Flap* or *Rheb* impairs tyrosine phosphorylation of STAT3, pharmacological inhibition of mTORC1 by rapamycin does not interfere with STAT3 phosphorylation.[2,4] Accordingly, deletion of *Raptor* impairs Th17 without affecting tyrosine phosphorylation of STAT3; whereas[2] deletion of *Raptor* does not inhibit Th1 differentiation. The differences between *Flap*, *Rheb*, and *Raptor* deficiencies in the regulation of STAT phosphorylation and Th1 differentiation suggest mTORC1-independent mechanisms are involved in the phosphorylation of STAT proteins by mTOR and Rheb.

We have also demonstrated that Th17 differentiation is impaired by mTORC1 inhibition via decreased RORγt nuclear translocation and increased *Gfi1* expression.[2] Expression of *Gfi1*, a negative regulator of Th17 differentiation, is suppressed by the transcription factor EGR2, the expression of which is positively regulated by S6K1 downstream of mTORC1.[2] In addition to suppressing *Gfi1* expression, mTORC1 accelerates the translocation of RORγt into nuclei during Th17 differentiation.[2] We have shown that S6K2, a nuclear counterpart of S6K1 that possesses a nuclear localization signal (NLS) in the C-terminus, associates with RORγt—which does not possess a NLS but is localized in the nucleus in Th17 cells—and transports RORγt to the nucleus. The amount of S6K2 protein increases upon TCR stimulation partly in a mTORC1-dependent manner; thus, the PI3K–Akt–mTORC1–S6K2 pathway also positively controls Th17 differentiation by enhancing nuclear translocation of RORγt.[2]

Conclusion

The differentiation of Th17 cells is controlled by a variety of intracellular signaling pathways and a complex transcription factor network including PI3K, Akt, and mTOR, as reviewed above. Signaling via mTORC2–Akt–FoxO1/3a has been well characterized to negatively regulate T_{reg} differentiation in a PI3K-dependent manner, but the contribution of this signaling pathway to Th17 differentiation is still controversial. PI3K–Akt–mTORC1 signaling positively regulates Th17 differentiation via multiple mechanisms including the regulation of HIF-1α expression, STAT3 phosphorylation, *Gfi1* downregulation, and the nuclear

translocation of RORγt. These accumulating pieces of evidence are providing more knowledge about T cell–mediated immunity and opportunities for manipulating immune systems in many inflammatory disorders.

Acknowledgments

This work was in part supported by a Grant-in-Aid for Young Scientist (B) (21790476 to S.N.), from the Japan Society for the Promotion of Science, a Grant-in-Aid from The Takeda Science Foundation (S.N.), a Grant-in-Aid from The Mochida Memorial Foundation for Medical and Pharmaceutical Research (S.N.), a Health Sciences Research Grant for Research on Specific Diseases from the Ministry of Health, Labour and Welfare, Japan, a Grant-in-Aid for Scientific Research on Priority Areas (20060021 to S.N.), a National Grant-in-Aid for the Establishment of a High-Tech Research Center in a Private University, a grant for the Promotion of the Advancement of Education and Research in Graduate Schools, and a Scientific Frontier Research Grant from the Ministry of Education, Culture, Sports, Science and Technology, Japan.

Conflicts of interest

S.K. is a consultant for Medical and Biological Laboratories, Co. Ltd. The authors otherwise have no financial conflicts of interest.

References

1. Ouyang, W., J.K. Kolls & Y. Zheng. 2008. The biological functions of T helper 17 cell effector cytokines in inflammation. *Immunity* **28:** 454–467.
2. Kurebayashi, Y., S. Nagai, A. Ikejiri, *et al.* 2012. PI3K-Akt-mTORC1-S6K1/2 axis controls Th17 differentiation by regulating *Gfi1* expression and nuclear translocation of RORg. *Cell Rep.* **1:** 360–373.
3. Okkenhaug, K., D.T. Patton, A. Bilancio, *et al.* 2006. The p110d isoform of phosphoinositide 3-kinase controls clonal expansion and differentiation of Th cells. *J. Immunol.* **177:** 5122–5128.
4. Lee, K., P. Gudapati, S. Dragovic, *et al.* 2010. Mammalian target of rapamycin protein complex 2 regulates differentiation of Th1 and Th2 cell subsets via distinct signaling pathways. *Immunity* **32:** 743–753.
5. Ouyang, W., O. Beckett, Q. Ma, *et al.* 2010. Foxo proteins cooperatively control the differentiation of Foxp3$^+$ regulatory T cells. *Nat. Immunol.* **11:** 618–627.
6. Harada, Y., C. Elly, G. Ying, *et al.* 2010. Transcription factors Foxo3a and Foxo1 couple the E3 ligase Cbl-b to the induction of Foxp3 expression in induced regulatory T cells. *J. Exp. Med.* **207:** 1381–1391.

7. Sauer, S., L. Bruno, A. Hertweck, *et al.* 2008. T cell receptor signaling controls Foxp3 expression via PI3K, Akt, and mTOR. *Proc. Natl. Acad. Sci. USA* **105:** 7797–7802.

8. Delgoffe, G.M., K.N. Pollizzi, A.T. Waickman, *et al.* 2011. The kinase mTOR regulates the differentiation of helper T cells through the selective activation of signaling by mTORC1 and mTORC2. *Nat. Immunol.* **12:** 295–303.

9. Ikejiri, A., S. Nagai, N. Goda, *et al.* 2011. Dynamic regula-

tion of Th17 differentiation by oxygen concentrations. *Int. Immunol.* **24:** 137–146.

10. Shi, L.Z., R. Wang, G. Huang, *et al.* 2011. HIF1a-dependent glycolytic pathway orchestrates a metabolic checkpoint for the differentiation of TH17 and Treg cells. *J. Exp. Med.* **208:** 1367–1376.

11. Dang, E.V., J. Barbi, H.Y. Yang, *et al.* 2011. Control of $T_H17/T(reg)$ balance by hypoxia-inducible factor 1. *Cell* **146:** 772–784.

Ann. N.Y. Acad. Sci. ISSN 0077-8923

ANNALS OF THE NEW YORK ACADEMY OF SCIENCES
Issue: *Inositol Phospholipid Signaling in Physiology and Disease*

Targeting phosphoinositide 3-kinase δ for the treatment of respiratory diseases

Srividya Sriskantharajah, Nicole Hamblin, Sally Worsley, Andrew R. Calver, Edith M. Hessel, and Augustin Amour

Refractory Respiratory Inflammation Discovery Performance Unit, Respiratory Therapy Area, GlaxoSmithKline, Stevenage, Hertfordshire, United Kingdom

Address for correspondence: Augustin Amour, Refractory Respiratory Inflammation Discovery Performance Unit, Respiratory Therapy Area, GlaxoSmithKline, Stevenage, SG1 2NY, Hertfordshire, UK. augustin.j.amour@gsk.com

Asthma and chronic obstructive pulmonary disease (COPD) are characterized in their pathogenesis by chronic inflammation in the airways. Phosphoinositide 3-kinase δ (PI3Kδ), a lipid kinase expressed predominantly in leukocytes, is thought to hold much promise as a therapeutic target for such inflammatory conditions. Of particular interest for the treatment of severe respiratory disease is the observation that inhibition of PI3Kδ may restore steroid effectiveness under conditions of oxidative stress. PI3Kδ inhibition may also prevent recruitment of inflammatory cells, including T lymphocytes and neutrophils, as well as the release of proinflammatory mediators, such as cytokines, chemokines, reactive oxygen species, and proteolytic enzymes. In addition, targeting the PI3Kδ pathway could reduce the incidence of pathogen-induced exacerbations by improving macrophage-mediated bacterial clearance. In this review, we discuss the potential and highlight the unknowns of targeting PI3Kδ for the treatment of respiratory disease, focusing on recent developments in the role of the PI3Kδ pathway in inflammatory cell types believed to be critical to the pathogenesis of COPD.

Keywords: respiratory inflammation; asthma; chronic obstructive pulmonary disease (COPD); T cell; neutrophil; epithelial cell; macrophage; phosphoinositide 3-kinase (PI3K)

Introduction

PI3Kδ is a lipid kinase involved in the activation, proliferation, and differentiation of leukocytes. It is being intensively investigated as a target for the treatment of inflammatory conditions, such as rheumatoid arthritis and asthma, as well as hematopoietic malignancies, including chronic lymphocytic leukemia and non-Hodgkin's lymphoma. The focus of our research has been to study the benefit of PI3Kδ inhibition in the context of obstructive lung diseases such as severe asthma and COPD. Chronic inflammation in the lung has long been linked to the pathogenesis of both these diseases, leading to chronic or intermittently compromised lung function and respiratory symptoms, such as cough, sputum production, wheeze, and breathlessness. We previously reported the potential benefits of targeting PI3Kδ for asthma,[1] and here we focus our attention on the potential for this target in the treatment of COPD.

COPD encompasses both bronchitis and emphysema, two conditions predominantly caused by smoking or air pollution.[2] Disease results from a chronic inflammatory process that causes excessive mucus production and remodeling of the upper airway tissues in bronchitis, and the destruction of alveolar walls in the lower airways in emphysema. Maintenance therapies for COPD are based on bronchodilators, inhaled corticosteroids, or combinations of both.[2] Although chronic, disease progression is marked by frequent episodes of acute inflammation in some patients, often following viral or bacterial infection of the airways. These exacerbation events are poorly controlled with current maintenance therapies, resulting in recurrent hospital admissions and the need for oral steroid treatment, which has many undesirable

doi: 10.1111/nyas.12039

side effects. There is therefore a significant unmet medical need in COPD for agents that can both deliver improved lung function and reduce exacerbations.[2]

Recent discoveries concerning the role of the PI3Kδ pathway in epithelial cells, macrophages, neutrophils, and T lymphocytes have generated increased interest in PI3Kδ as a target for COPD (summarized in Fig. 1), and in this review we highlight some of these key findings that we believe reflect current thinking in the field.

Epithelial cells

Airborne irritants such as cigarette smoke activate epithelial cells to release inflammatory mediators, a process thought to initiate the cycle of chronic inflammation in COPD.[2] There is evidence that the PI3Kδ signaling pathway may play an important role in airway epithelial cells. For instance, in tracheal epithelial cells taken from mice challenged with the allergen ovalbumin, a selective PI3Kδ inhibitor reduced the expression of VEGF, a mediator of lung vascular permeability in respiratory diseases.[3] This was rather unexpected since PI3Kδ is not the most abundantly expressed PI3K isoform in epithelial cells. However, PI3Kδ expression or activity may increase in inflammatory conditions, as seen with endothelial cells where TNFα upregulates PI3Kδ expression.[4] More recently, the cytokine IL-17 has been shown to induce loss of HDAC2 and steroid sensitivity in a bronchial epithelial 16HBE cell line, which can be reversed by a nonselective PI3K inhibitor.[5] Since similar observations have been made in the context of oxidative stress in macrophages and have further specifically implicated PI3Kδ,[2] it would be interesting to determine if PI3Kδ is also involved in the IL-17–evoked steroid insensitivity in primary human bronchial epithelial cells. Oxidation is known to inhibit tyrosine phosphatases and will therefore indirectly activate class IA PI3Ks such as PI3Kδ. Interestingly, the binding affinity of the COPD drug theophylline for PI3Kδ is improved by two orders of magnitude when A549 cells are pretreated with hydrogen peroxide.[2] Therefore conditions of oxidative stress may induce the direct activation and conformational change of PI3Kδ in epithelial cells. Overall, however, the regulation of PI3Kδ expression and activity by inflammatory mediators or oxidative stress remains poorly understood in airway epithelial cells.

Macrophages

Macrophages are rapidly emerging as major mediators of inflammation in COPD, releasing proteases and chemokines that attract T cells, neutrophils, and monocytes to the lung. Lungs of COPD patients show an approximate 25-fold increase in macrophage number compared to lungs from healthy individuals.[2] Since β1- and β2-integrin dependent monocyte adhesion and migration require PI3Kδ, its inhibition should impair the increased monocyte infiltration observed in COPD.[6] Expression of PI3Kδ and Akt phosphorylation are also enhanced in COPD alveolar macrophages (AM), and oxidative stress directly induces PI3Kδ-dependent Akt activation, leading to a loss of steroid sensitivity.[2] This observation has led to the hypothesis that a PI3Kδ inhibitor may complement steroids in conditions of oxidative stress and could therefore improve the clinical response to steroids as an adjunct therapy for COPD. BAL macrophages from COPD patients have also been found to be more M2-like than the predominantly classical M1 macrophages of healthy nonsmokers.[7] A skewing toward the tissue remodeling function of M2 genes may therefore contribute to the fibrosis and poor pathogen clearance seen in COPD. Interestingly, alternative M2 activation of macrophages is sensitive to PI3Kδ inhibition in IL-4–stimulated murine macrophages.[8] Taken together, these observations suggest that several facets of macrophage function implicated in COPD pathology may be targeted by PI3Kδ inhibition. However, the intricacies of PI3Kδ function in COPD macrophages require further investigation.

Neutrophils

Neutrophils are regarded as key mediators of inflammation in COPD, as increased presence in sputum of patients correlates with disease severity.[2] Since neutrophil migration relies on PI(3,4,5)P$_3$ accumulation in the leading edge, there is strong evidence for PI3K involvement in chemotaxis of neutrophils. Furthermore, the faster and less accurate chemotaxis ascribed to COPD neutrophils can be reversed by treatment with a pan-PI3K inhibitor.[9] Although the specific role of PI3Kδ in COPD neutrophil migration remains to be explored, it likely plays a major role, as PI3Kδ is necessary for directional migration to formyl-methionyl-leucyl-phenylalanine

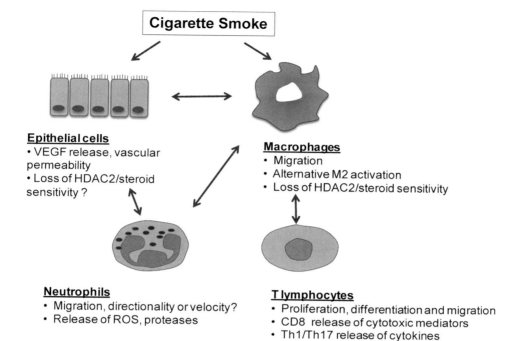

Figure 1. The key cell types involved in the inflammatory process of COPD and their features that are predicted to be modulated by a PI3Kδ inhibitor.

(fMLP) in healthy neutrophils. Moreover, PI3Kδ inhibition restricts *in vivo* neutrophil migration into inflamed tissue in a murine LPS model of acute lung injury.[10] Notably, this inhibition of neutrophil influx is partial (60–80% reduction), which may be beneficial in retaining the defense neutrophils provide against pathogens during COPD exacerbations.

Neutrophil activation leads to respiratory burst and the release of enzymes that contribute to lung tissue damage in COPD. Indeed, neutrophil elastase (NE) and matrix metalloproteinase-9 (MMP-9) activity increase with disease severity.[2] Release of NE and MMP-9 induced by fMLP from healthy blood neutrophils can be reduced significantly, but not completely, by pan-PI3K treatment *in vitro*.[11] Since this release is insensitive to steroid treatment, PI3K inhibitors may confer an additional benefit to steroids by reducing neutrophilic inflammation in COPD patients. This effect presumably involves PI3Kδ, since a separate study has shown that TNF-α- and fMLP-induced NE and superoxide release are significantly reduced by PI3Kδ selective inhibition.[12] Interestingly, *in vivo* NE release seems to be more sensitive to PI3Kδ inhibition than neutrophil migration in a TNF-α-induced air pouch model.[12]

However, further work is required to understand how PI3Kδ inhibition may alter neutrophil migration and degranulation in the context of COPD.

T cells

T lymphocytes are important effectors and regulatory cells that accumulate in lung tissue and airways of COPD patients. PI3Kδ is activated by T cell receptor and tyrosine kinase cytokine receptor signalling and is therefore central to T cell function.[13] CD8[+] T cells, mainly the IFN-γ–and GMCSF-secreting Tc1 subset, have been linked to COPD pathology, since their number in the airways of COPD patients positively correlates with disease severity.[2] Since PI3Kδ is required for TCR-induced IFN-γ production and the proliferation of CD8[+] T cells,[13] PI3Kδ inhibition may reduce their number and impact in COPD lungs. Considering that selective PI3Kδ inhibition increases trafficking of CD8[+] cells to lymph nodes, it may also reduce the excessive CD8[+] cell migration into the lungs of COPD patients. Additionally, enhanced CD8[+]-mediated cytotoxicity is thought to contribute to the increased apoptosis of pneumocytes observed in emphysema, PI3Kδ inhibition is likely to reduce this process by preventing perforin and granzyme expression.[14] Curiously, despite enhanced cytotoxicity of CD8[+] T cells, viral infections persist during COPD exacerbations. This suggests that misdirected and dysregulated CD8[+] response occurs in COPD, and the effect of a PI3Kδ inhibitor in this context remains intriguingly in question.

In addition, COPD is considered to have an autoimmune element with CD4[+] Th1 and Th17 cells playing central roles.[2] PI3Kδ inhibition is known to reduce the differentiation and proliferation of Th1 and Th17 cells, as well as the cytokines IFN-γ and IL-17 they respectively secrete.[13] A PI3Kδ inhibitor is thus likely to profoundly dampen the autoimmune component of COPD. The role of the suppressive T_{reg} cell subset in COPD is less clear, with both an increase and decrease in airway number described. Peripheral maintenance of murine T_{reg} cells requires PI3Kδ activity, and $p110\delta^{D910A}$ T_{reg} cells are less suppressive than their wild-type counterparts.[13] Although loss of T_{reg} cell function is generally undesirable, under some conditions it may be protective by enhancing immune responses. Indeed, $p110\delta^{D910A}$ mice develop spontaneous colitis but are resistant to *Leishmania major* infection as a result of unrestrained macrophage function.[15] Taken together, descriptive evidence suggests that PI3Kδ inhibition can reset the balance of CD8[+] and CD4[+] T cell function that may be beneficial in COPD. Overall, more direct evidence examining the phenotype of the altered lymphocytes from COPD patients as well as the impact of PI3Kδ inhibitors on these cells is required.

Conclusions

PI3Kδ has emerged as a potentially important therapeutic target for the treatment of COPD. It is likely to perform key functions in inflammatory cells that have been shown to contribute to COPD pathology. However, COPD is a complex condition with a poorly defined mechanism of pathogenesis, and therefore the development of new therapeutic approaches to tackle COPD requires a better understanding of the altered lung environment, cell infiltration, cellular function, and signaling pathways in the disease state. Nevertheless, the findings reviewed above offer much promise that PI3Kδ

inhibition could reduce inflammation, improve pathogen clearance during exacerbations, and restore steroid sensitivity in COPD. Overall, these observations suggest that further investigation into the potential of PI3Kδ inhibition in COPD is warranted.

Conflicts of interest

We are all GSK employes and, as such, are also GSK shareholders.

References

1. Rowan, W.C., J.L. Smith, K. Affleck & A. Amour. 2012. Targeting phosphoinositide 3-kinase δ for allergic asthma. *Biochem. Soc. Trans.* **40:** 240–245.
2. Barnes, P. 2012. "Inflammatory and immune mechanisms in COPD." In *Advances in Chronic Obstructive Pulmonary Disease (COPD): New Mechanisms, Management Strategies and Treatments, The Biomedical & Life Sciences Collection.* P. Barnes, Ed. Henry Stewart Talks Ltd. London. Available at: http://hstalks.com/bio.
3. Kim, S.R., K.S. Lee, H.S. Park, *et al.* 2010. HIF-1α inhibition ameliorates an allergic airway disease via VEGF suppression in bronchial epithelium. *Eur. J. Immunol.* **40:** 2858–2869.
4. Whitehead, M.A., M. Bombardieri, C. Pitzalis & B. Vanhaesebroeck. 2012. Isoform-selective induction of human p110δ PI3K expression by TNFα: identification of a new and inducible PIK3CD promoter. *Biochem. J.* **443:** 857–867.
5. Zijlstra, G.J., N.H. Ten Hacken, R.F. Hoffmann, *et al.* 2012. Interleukin-17A induces glucocorticoid insensitivity in human bronchial epithelial cells. *Eur. Respir. J.* 39: 439–445.
6. Ferreira, A.M., H. Isaacs, J.S. Hayflick, *et al.* 2006. The p110δ isoform of PI3K differentially regulates β1 and β2 integrin-mediated monocyte adhesion and spreading and modulates diapedesis. *Microcirculation* **13:** 439–456.
7. Shaykhiev, R., A. Krause, J. Salit, *et al.* 2009. Smoking-dependent reprogramming of alveolar macrophage polarization: implication for pathogenesis of chronic obstructive pulmonary disease. *JI* 183: 2867–2883.
8. Weisser, S.B., K.W. McLarren, N. Voglmaier, *et al.* 2011. Alternative activation of macrophages by IL-4 requires SHIP degradation. *Eur. J. Immunol.* **41:** 1742–1753.
9. Sapey, E., J.A. Stockley, H. Greenwood, et al. 2011. Behavioral and structural differences in migrating peripheral neutrophils from patients with chronic obstructive pulmonary disease. *Am. J. Respir. Crit. Care Med.* **183:** 1176–1186.
10. Puri, K.D., T.A. Doggett, J. Douangpanya, *et al.* 2004. Mechanisms and implications of phosphoinositide 3-kinase delta in promoting neutrophil trafficking into inflamed tissue. *Blood* 103: 3448–3456.
11. Vlahos, R., P.A. Wark, G.P. Anderson & S. Bozinovski. 2012. Glucocorticosteroids differentially regulate MMP-9 and neutrophil elastase in COPD. *PLoS One* **7:** e33277.

12. Sadhu, C., K. Dick, W.T. Tino & D.E. Staunton. 2003. Selective role of PI3K delta in neutrophil inflammatory responses. *Biochem. Biophys. Res. Commun.* **308:** 764–769.

13. So, L. & D.A. Fruman. 2012. PI3K signalling in B- and T-lymphocytes: new developments and therapeutic advances. *Biochem. J.* **442:** 465–481.

14. Macintyre, A.N., D. Finlay, G. Preston, *et al.* 2011. Protein kinase B controls transcriptional programs that direct cytotoxic T cell fate but is dispensable for T cell metabolism. *Immunity* **34:** 224–236.

15. Liu, D., T. Zhang, A.J. Marshall, *et al.* 2009. The p110delta isoform of phosphatidylinositol 3-kinase controls susceptibility to Leishmania major by regulating expansion and tissue homing of regulatory T cells. *J. Immunol.* **183:** 1921–1933.

Ann. N.Y. Acad. Sci. ISSN 0077-8923

ANNALS OF THE NEW YORK ACADEMY OF SCIENCES

Issue: *Inositol Phospholipid Signaling in Physiology and Disease*

Phosphoinositide 3-kinase gamma in T cell biology and disease therapy

Wai-Ping Fung-Leung

Janssen Research & Development, L.L.C., San Diego, California

Address for correspondence: Wai-Ping Fung-Leung, Janssen Research & Development, L.L.C., 3210 Merryfield Row, San Diego, CA 92121. wleung@its.jnj.com

Phosphoinositide 3-kinase gamma (PI3Kγ) kinase activity is important for its signaling functions in T cell development, activation, differentiation, and trafficking. Protection of PI3Kγ knockout mice from disease in multiple autoimmune models suggests that targeting PI3Kγ alone, or in combination with PI3Kδ, could be a promising approach to disease therapy.

Keywords: PI3Kγ; T cells; disease therapy

Introduction

Phosphoinositide 3-kinase gamma (PI3Kγ) is a lipid kinase that generates phosphatidylinositol 3,4,5-trisphosphate (PIP$_3$) to recruit and activate downstream signaling molecules. PI3Kγ is expressed mainly in cells of hematopoietic lineage, suggesting that it has a specific role in the immune system. It is a heterodimer consisting of a catalytic subunit p110γ and one of the two regulatory subunits p101 and p84. Although the two regulatory subunits have a common role in regulating PI3Kγ kinase activity, they exhibit distinct modulatory effects on PI3Kγ signaling that lead to different biological outcomes. PI3Kγ is activated by G protein–coupled receptors (GPCRs) through interaction with G$_{\beta\gamma}$ proteins. Kinase activity is believed to be the key mechanism in PI3Kγ signaling, although its role as a scaffolding protein has also been demonstrated in the cardiovascular system.

PI3Kγ in T cell development

The role of PI3Kγ in T cell development has been characterized in PI3Kγ knockout mice.[1,2] The lack of PI3Kγ expression results in defective T cell receptor (TCR) β chain selection at the CD4/CD8 double negative DN3 stage of development, that leads to a decrease in double positive (DP) thymocytes and peripheral T cells. Mice defective in PI3Kγ kinase activity have been developed by introduction of a point mutation in the p110γ gene;[3] a similar defect is observed in PI3Kγ kinase-inactive thymocytes (unpublished data). The published results suggest that PI3Kγ enzymatic activity plays a critical role in thymic T cell development. PI3Kγ is not the only PI3K member participating in thymic development, however. Mice lacking the expression of both PI3Kγ and PI3Kδ exhibit severe thymic atrophy and peripheral lymphopenia, suggesting that these PI3K members display some nonredundant roles in thymocyte development.[4] It is well recognized that the major function of thymic development is positive and negative selection of thymocytes to generate T cells with a highly diverse TCR repertoire for immune surveillance, while keeping autoreactivity in check. Thymic development of T cells expressing transgenic TCRs is significantly reduced in PI3Kγ knockout mice, suggesting a potential defect in thymic positive selection.[5] However, detailed studies are still needed to fully understand the role of PI3Kγ in thymic selection.

PI3Kγ in T cell activation

The role of PI3Kγ in T cell activation remains controversial, partly due to the subtle defects observed in PI3Kγ knockout T cells and the limited information on PI3Kγ in TCR signaling. Studies

doi: 10.1111/nyas.12029

 Ann. N.Y. Acad. Sci. 1280 (2013) 40–43 © 2013 New York Academy of Sciences.

with PI3Kγ knockout T cells have found reduced proliferation and cytokine production when the cells are stimulated with anti-CD3 alone or in combination with anti-CD28.[1,6] The typical defect is a partial blockade of T cell activation that cannot be rescued with exogenous IL-2. Interestingly, T cells from PI3Kδ kinase-inactive mice show similar subtle defects. To confirm the role of PI3Kγ in T cell activation and to verify the importance of its kinase activity, we studied T cells from PI3Kγ kinase-inactive mice with the same approach. A mild reduction in T cell proliferation and cytokine production was observed, and the remaining activity could be completely blocked by PI3Kδ inhibitors; these defects were recapitulated with antigen-specific stimulation (unpublished data). It seems that both PI3Kγ and PI3Kδ contribute to T cell activation via their kinase activities. PI3Kδ probably plays a stronger role than PI3Kγ, but a more significant blocking effect is achieved when the function of both is inhibited.

PI3Kγ in TCR signaling

When TCR is engaged with antigenic peptide on MHC molecules, a cascade of signaling events is initiated that eventually leads to cell proliferation and gene induction. TCR signaling is induced by tyrosine kinases Lck and Fyn, which phosphorylate the cytoplasmic portion of the CD3 complex to provide docking sites for ZAP70. ZAP70 subsequently phosphorylates adaptor proteins such as LAT to recruit a number of signaling proteins, including PLCγ1, to the TCR complex. PLCγ1 hydrolyzes phosphatidylinositol 4,5-diphosphate (PIP_2) to yield two lipid messengers inositol 1,4,5-triphosphate (IP_3) and diacylglycerol (DAG). DAG recruits Ras and PKC-θ to the membrane to initiate the Ras–Erk and NF-κB pathways, whereas IP_3 triggers calcium depletion from internal stores, which leads to extracellular calcium influx and initiation of the NFAT signaling pathway. Class IA PI3Ks are known to interact with the TCR complex by binding to Lck and the G proteins Rac and Ras. Costimulatory molecules CD28 and ICOS also have class IA PI3K binding motifs (YXXM) on their cytoplasmic tails.

In contrast, the interaction of PI3Kγ with TCR complex remains unclear. Although PI3Kγ has been shown to associate with $G\alpha_{q/11}$, Lck, and ZAP70,[6] details on its involvement in TCR signaling require further investigation. Chemokine receptors on T cells such as CXCR4 have been suggested to function

as TCR costimulators. CXCR4 is recruited to the immune synapse and associates with the TCR signaling complex during T cell activation. Interestingly, CXCR4 is also required for TCR β chain selection during thymic development. Finally, PI3Kγ is known to be activated by GPCRs, and it is possible that its involvement in TCR signaling is through participation in signaling events of chemokine receptors that function as costimulators.

PI3Kγ in T cell differentiation

Successful T cell activation allows the expansion and polarization of T cells into effector cells equipped with specific functions. It has been demonstrated that differentiation of $CD4^+$ T cells into Th1, Th2, Th17, and regulatory T cells is guided by specific cytokines. For example, IL-12 and IFN-γ are the key driving cytokines for Th1 cells, whereas IL-4 is essential for Th2 cell polarization. In addition, the generation of regulatory T cells with TGF-β1 can be redirected to Th17 cells by IL-23, IL-6, and IL-1β. Th17 cells can be pathogenic in autoimmune diseases, whereas regulatory T cells are important for immune regulation.

PI3K has been shown to play a role in T cell differentiation by modulating activities of transcription factors FoxO and KLF2.[7] Polarization to all subsets of T cells, including regulatory T cells, is defective in PI3Kδ kinase-inactive mice.[8,9] Human $CCR6^+$ Th17 memory T cells produce high levels of IL-17A when stimulated with IL-7, and this induction is blocked by PI3K inhibitors.[7]

Recently a PI3Kγ inhibitor was reported to block Th17 differentiation in human CD4 T cells.[10] PI3Kγ-deficient mice were also shown to be protected in a model of Th17 cell–driven psoriasis.[11] Future studies to define the mechanism and the specific PI3K members required for T cell differentiation will be valuable for the design of novel therapeutic interventions for disease treatment.

PI3Kγ in T cell trafficking

PI3Kγ is the key PI3K member that interacts with $G_{\beta\gamma}$ proteins downstream of GPCR and mediates the leukocyte chemotactic response. Chemokine receptors, such as CXCR4, play an important role in directing thymocyte migration through the cortex and medulla during T cell development. Proper T cell trafficking in response to infections is also tightly regulated by chemokines. CCR7 is one of the chemokines needed for homing of naive T cells to lymph nodes, where they are activated by antigen presenting cells and differentiate into effector cells. Effector T cells respond to inflammation signals and infiltrate peripheral tissues to exert their effector functions in pathogen clearance.

PI3Kγ-knockout T cells are defective in chemotaxis toward a number of chemokines.[12,13] Similar defects have been observed in T cells from PI3Kγ kinase-inactive mice (unpublished data); we have also observed a reduction in the T cell–mediated delayed-type hypersensitivity response in PI3Kγ kinase-inactive mice (unpublished data).

PI3Kγ in drug discovery and disease therapy

PI3Kγ is a target of interest in drug discovery for treatment of inflammation and autoimmune diseases. PI3Kγ-knockout mice are protected in a number of disease models, including arthritis, psoriasis, lupus, colitis, experimental autoimmune encephalomyelitis, atherosclerosis, and asthmatic models. The positive phenotype demonstrated in these animal models is likely due to multiple mechanisms on more than one leukocyte cell type that are the result of PI3Kγ deficiency. Inhibition of PI3Kγ kinase activity, but not of its adaptor function, could be an ideal scenario in drug discovery to achieve efficacy while avoiding cardiovascular side effects. It is therefore imperative for both basic research and drug discovery to understand the molecular mechanism of PI3Kγ signaling and biology. PI3Kγ kinase-inactive mice have proved to be a valuable animal model with which to define its biological role. The X-ray crystallography structure of PI3Kγ has been elucidated and reveals the key molecular interactions of PI3Kγ with small molecule inhibitors.[14] A number of PI3Kγ inhibitors have been generated and their efficacies have been demonstrated in different biological systems.[15] However, caution should be taken in interpreting the results from compounds with limited target selectivity. Compound selectivity, as demonstrated in kinase assays, may not necessarily be translated to an intracellular environment with high levels of ATP; for example, compound potencies in cells may decrease significantly if their kinases have high ATP binding affinities. Indeed, PI3Kγ has the highest ATP binding affinity among PI3K class I members and thus it is likely to be the most challenging member for targeted inhibition.

Recently, there has been increased interest in developing dual PI3Kγ/δ inhibitors that could achieve better therapeutic efficacy while maintaining acceptable safety profiles. Indeed, enhanced protection from disease progression has been demonstrated in arthritic models when both PI3Kγ and PI3Kδ were inactivated by gene targeting or inhibitors.[16] Orally available medicines could be a more favorable treatment for chronic autoimmune disease despite recent advances in biologic therapies. Targeting PI3Kγ and PI3Kδ, individually or in combination, with small molecule inhibitors may be a promising treatment for autoimmune diseases.

Acknowledgments

PI3Kγ kinase-inactive mice were kindly provided by Emilio Hirsch from University of Torino. Nadia Ladygina, a postdoctoral fellow at Janssen R&D, contributed to our studies with PI3Kγ kinase-inactive mice.

Conflicts of interest

The author is an employee of Janssen R&D.

References

1. Sasaki, T. *et al.* 2000. Function of PI3Kgamma in thymocyte development, T cell activation, and neutrophil migration. *Science* **287**: 1040–1046.
2. Rodriguez-Borlado, L. *et al.* 2003. Phosphatidylinositol 3-kinase regulates the CD4/CD8 T cell differentiation ratio. *J. Immunol.* **170**: 4475–4482.
3. Patrucco, E. *et al.* 2004. PI3Kgamma modulates the cardiac response to chronic pressure overload by distinct kinase-dependent and -independent effects. *Cell* **118**: 375–387.
4. Webb, L.M. *et al.* 2005. Cutting edge: T cell development requires the combined activities of the p110gamma and p110delta catalytic isoforms of phosphatidylinositol 3-kinase. *J. Immunol.* **175**: 2783–2787.
5. Rodrigues, D.H. *et al.* 2006. Investigation of nociception and leukocyte recruitment in an experimental model of multiple sclerosis in a PI3 kinase-deficient mice. XXXI Meeting of Brazilian Society for Immunology.
6. Alcazar, I. *et al.* 2007. Phosphoinositide 3-kinase gamma participates in T cell receptor-induced T cell activation. *J. Exp. Med.* **204**: 2977–2987.
7. Wan, Q. *et al.* 2011. Cytokine signals through PI-3 kinase pathway modulate Th17 cytokine production by CCR6+human memory T cells. *J. Exp. Med.* **208**: 1875–1887. Epub 2011 Aug 8.

8. Okkenhaug, K. *et al.* 2006. The p110delta isoform of phosphoinositide 3-kinase controls clonal expansion and differentiation of Th cells. *J. Immunol.* **177:** 5122–5128.

9. Patton, D.T. *et al.* 2006. Cutting edge: the phosphoinositide 3-kinase p110 delta is critical for the function of CD4+CD25+Foxp3+ regulatory T cells. *J. Immunol.* **177:** 6598–6602.

10. Bergamini, G. *et al.* 2012. A selective inhibitor reveals PI3Kgamma dependence of T(H)17 cell differentiation. *Nat. Chem. Biol.* **8:** 576–582.

11. Roller, A. *et al.* 2012. Blockade of phosphatidylinositol 3-Kinase (PI3K)delta or PI3Kgamma reduces IL-17 and ameliorates imiquimod-induced psoriasis-like dermatitis. *J. Immunol.* **189:** 4612–4620.

12. Reif, K. *et al.* 2004. Cutting edge: differential roles for phosphoinositide 3-kinases, p110gamma and p110delta, in lymphocyte chemotaxis and homing. *J. Immunol.* **173:** 2236–2240.

13. Martin, A.L. *et al.* 2008. Selective regulation of CD8 effector T cell migration by the p110 gamma isoform of phosphatidylinositol 3-kinase. *J. Immunol.* **180:** 2081–2088.

14. Zvelebil, M.J., M.D. Waterfield & S.J. Shuttleworth. 2008. Structural analysis of PI3-kinase isoforms: identification of residues enabling selective inhibition by small molecule ATP-competitive inhibitors. *Arch. Biochem. Biophys.* **477:** 404–410.

15. Venable, J.D. *et al.* 2010. Phosphoinositide 3-kinase gamma (PI3Kgamma) inhibitors for the treatment of inflammation and autoimmune disease. *Recent Pat. Inflamm. Allergy Drug Discov.* **4:** 1–15.

16. Randis, T.M. *et al.* 2008. Role of PI3Kdelta and PI3Kgamma in inflammatory arthritis and tissue localization of neutrophils. *Eur. J. Immunol.* **38:** 1215–1224.

Ann. N.Y. Acad. Sci. ISSN 0077-8923

ANNALS OF THE NEW YORK ACADEMY OF SCIENCES
Issue: *Inositol Phospholipid Signaling in Physiology and Disease*

Inhibition of phosphoinositide 3-kinase γ attenuates inflammation, obesity, and cardiovascular risk factors

Matthias P. Wymann[1] and Giovanni Solinas[2]

[1]Department of Biomedicine, University of Basel, Basel, Switzerland. [2]Department of Medicine, University of Fribourg, Fribourg, Switzerland

Address for correspondence: Matthias P. Wymann, Department of Biomedicine, University of Basel, Mattenstrasse 28, 4058 Basel, Switzerland. Matthias.Wymann@UniBas.CH

Phosphoinositide 3-kinase γ (PI3Kγ) plays a central role in inflammation, allergy, cardiovascular, and metabolic disease. Obesity is accompanied by chronic, low-grade inflammation. As PI3Kγ plays a major role in leukocyte recruitment, targeting of PI3Kγ has been considered to be a strategy for attenuating progression of obesity to insulin resistance and type 2 diabetes. Indeed, PI3Kγ null mice are protected from high fat diet–induced obesity, metabolic inflammation, fatty liver, and insulin resistance. The lean phenotype of the PI3Kγ-null mice has been linked to increased thermogenesis and energy expenditure. Surprisingly, the increase in fat mass and metabolic aberrations were not linked to PI3Kγ activity in the hematopoietic compartment. Thermogenesis and oxygen consumption are modulated by PI3Kγ lipid kinase–dependent and –independent signaling mechanisms. PI3Kγ signaling controls metabolic and inflammatory stress, and may provide an entry point for therapeutic strategies in metabolic disease, inflammation, and cardiovascular disease.

Keywords: phosphoinositide 3-kinase; G protein–coupled receptors; chronic inflammation; obesity; atherosclerosis; leukocytes; adipocytes; thrombocytes; plaque formation

Introduction

Phosphoinositide 3-kinases (PI3Ks) have been shown to play key roles in the control of cellular metabolism, cell growth, proliferation, survival, and migration. The deregulation of these processes promotes chronic inflammation, tumor progression, and metabolic deviations (for a review, see Ref. 1). Class I PI3Ks produce the lipid second messenger phosphoinositol(3,4,5)-trisphosphate (PtdIns(3,4,5)P$_3$), which serves as a docking site for selected pleckstrin homology (PH) domain–containing proteins, including phosphoinositide-dependent kinase 1 (PDK1) and protein kinase B (PKB/Akt). While class IA PI3Ks are tightly associated with a p85-like regulatory subunit (encoded by the *PIK3R1*, *PIK3R2*, and *PIK3R3* gene loci) and are linked to protein tyrosine kinase receptor activation, the only member of the class IB PI3Ks, PI3Kγ, forms a heterodimer with a p84 or p101 adaptor protein. PI3Kγ operates downstream of G protein–coupled receptors (GPCRs)[2,3] and is activated by trimeric G protein G$_{\beta\gamma}$ subunits. Both the p84 and p101 adaptor proteins sensitize the catalytic subunit of PI3Kγ (p110γ) for G$_{\beta\gamma}$ input,[4] but the p110γ–p84 complex additionally requires interaction with the activated small G protein Ras to be fully operational.[5] The p110γ–p84 and p110γ–p101 complexes can signal in a nonredundant fashion, and can operate in distinct plasma membrane micro-domains.[3]

PI3Kγ promotes inflammation and modulates cardiovascular parameters

It was demonstrated early on that PI3Kγ attenuates *in vivo* leukocyte migration toward chemokines: neutrophil recruitment[2] and dendric cell movement[6] are attenuated in mice lacking a functional p110γ subunit. Protection of 110γ null animals was shown to involve control of the replenishment of inflammatory cells to inflammation sites in inflammatory and allergic disease models. This observation was also observed following selective PI3Kγ

doi: 10.1111/nyas.12037

Ann. N.Y. Acad. Sci. 1280 (2013) 44–47 © 2013 New York Academy of Sciences.

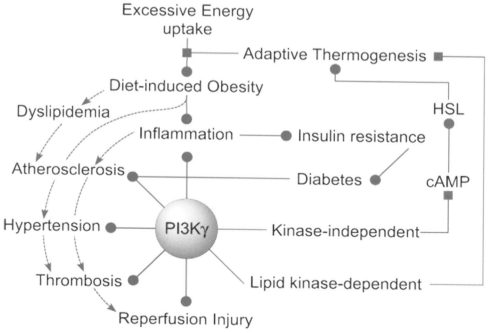

Figure 1. Implications of PI3Kγ signaling in metabolic disease, obesity, and cardiovascular disease. PI3Kγ downstream signaling attenuates adaptive thermogenesis via a kinase-independent and a lipid kinase–dependent pathway. Mice with functional PI3Kγ therefore accumulate more body fat when kept on a high fat diet. The onset of obesity promotes the development of metabolic disease, resulting in dyslipidemia, metabolic inflammation, and insulin-resistant tissues. An obesity-dependent, persistent low-grade inflammatory state contributes to macrophage recruitment to adipose tissue. Furthermore, PI3Kγ drives macrophage accumulation in the intima, and thus contributes to atherosclerotic plaque formation. Hypertension, linked to PI3K through the action of angiotensin II, favors rupture of atherosclerotic plaques. Finally, thrombosis can lead to blockage of coronary arteries. After a cardiac event, PI3Kγ can support inflammatory processes worsening reperfusion injury. HSL: hormone sensitive lipase.

inhibition. It was subsequently demonstrated that inhibition of PI3Kγ alleviated symptoms in rheumatoid arthritis[7] and systemic lupus mouse models,[8] and that anaphylaxis was promoted by mast cell activation and degranulation driven by a PI3Kγ-dependent autocrine/paracrine activation loop.[9]

Furthermore, in PI3γ-null platelets, aggregation and thrombosis were reduced in response to ADP. Loss of PI3Kγ activity correlates with attenuated $\alpha(\text{IIb})\beta(3)$ fibrinogen receptor activation-and protects from ADP-induced platelet-dependent thromboembolic vascular occlusion.[10] While cells of hematopoietic origin express high levels of p110γ protein, PI3Kγ also plays an important role in tissues where it is less prominently expressed. For example, in cardiomyocytes p110γ is linked to cardiac contractility, serving as a scaffold protein for phosphodiesterase 3B (PDB3). Lack of p110γ leads to increased basal and adrenergic receptor–

induced cAMP-protein kinase A (PKA) signaling in cardiomyocytes, and this translates into increased heart contractile force. Mice expressing a catalytically inactive (KI) p110γ (with Lys833 mutated to Arg), mimicking pharmaceutical inhibition of the enzyme, show no increase in cAMP, demonstrating that this pathway operates independently of PI3Kγ lipid kinase activity. Moreover, p110[KI/KI] mice were protected in a model of cardiac overload.[11]

In atherosclerosis, the role of inflammation has attracted attention, as some steps in the action of atherogenic lipoproteins and cytokines act through PI3K. PI3Kγ seems to promote early steps in the generation of atherosclerotic lesions. In murine models of atherosclerosis, such as the apolipoprotein E (ApoE)-null and low density lipoprotein receptor (LDLR)-null mice, either genetic (ApoE[−/−] × p110γ[−/−] or LDLR[−/−] × p110γ[−/−]) or pharmacological manipulation of PI3Kγ activity counteracted plaque formation. Interestingly, plaques forming in p110γ-null mice were stabilized by the inclusion of increased levels of collagen, which clinically correlates with a better prognosis. Plaque formation was also linked to the hematopoietic cell lineages, as the transplantation of p110γ$^{-/-}$ bone marrow into a wild-type host recapitulated the p110γ-null phenotype. Mechanistically, loss of p110γ decreases macrophage and T cell infiltration into the intima.[12,13] Once the atherosclerotic lesions progress to narrowing of blood vessels through stenosis, smooth muscle migration has been reported to be potentiated by PI3Kγ-dependent signals.[14]

As atherosclerotic plaques are formed, high blood pressure is a major risk factor for subsequent plaque rupture, stenosis, and, finally, fatal thrombotic events. In the development of hypertension, the renin–angiotensin system plays a major role by generating angiotensin II that elicits reactive oxygen species (ROS) within endothelial cells via GPCR stimulation. ROS production and the opening of PI3Kγ-dependent L-type Ca^{2+} channels, leading to influx of extracellular Ca^{2+}, mediate smooth muscle contraction. Most importantly, mice lacking PI3Kγ are protected from angiotensin II–induced hypertension, ROS production, and vascular damage.[15]

PI3Kγ–insulin resistance, control of thermogenesis, and obesity

Obesity is accompanied by chronic low-grade inflammation (metabolic inflammation), which has been proposed to be a cause of progressing insulin resistance, leading to the initiation of type II diabetes in obese patients.[16,17] It was also proposed that metabolic inflammation affects energy balance during the development of obesity.[18] Indeed, clinical studies suggest that specific anti-inflammatory treatments can improve glucose homeostasis in diabetics. Along these lines, we and others have recently found that loss of functional PI3Kγ leads to a major improvement of insulin sensitivity in mice kept on a high fat diet. Obesity-dependent macrophage infiltration into adipose tissue was attenuated in p110γ-null animals, and macrophage markers and inflammatory cytokine profiles were reduced in white adipose tissue.[19,20] One interesting outcome of our study was the observation that p110γ$^{-/-}$ mice on a high fat diet accumulated substantially less fat mass than wild-type mice, while calorie intake and nonadipose tissue mass was unaffected. The difference in body weight increase could be linked to increased thermogenesis in p110γ-null animals, triggered by lipid kinase–dependent and –independent pathways. Moreover, the lean phenotype accompanying increased thermogenesis in p110γ-null mice was independent from PI3Kγ activity within the hematopoietic compartment, since the energy expenditure and oxygen consumption was determined by the PI3Kγ status of the host and not by the genotype of the transplanted bone marrow.[20]

The PI3Kγ signaling signature reflects a role in control of metabolic and inflammatory stress. As described above and summarized in Figure 1, PI3Kγ inhibition might constitute the entry point for strategies for therapeutic intervention in metabolic and inflammatory disease. Additionally, multiple connections to PI3Kγ signaling involve processes in atherosclerotic plaque formation, platelet aggregation, hypertension, and others, providing novel possibilities for cardioprotective therapies.

Acknowledgments

We are grateful for support from the Swiss National Science Foundation Grants 310030–127574 and 31EM30–126143 (to M.P.W.), 31003A-118172 and 31003A˙135684 (to G.S.), and from the European Foundation for the Study of Diabetes (EFSD).

Conflicts of interest

The authors declare no conflicts of interest.

References

1. Wymann, M.P., R. Schneiter 2008. Lipid signalling in disease. *Nat. Rev. Mol. Cell Biol.* **9:** 162–176.
2. Hirsch, E., V.L. Katanaev, C. Garlanda, *et al*. 2000. Central role for G protein-coupled phosphoinositide 3-kinase gamma in inflammation. *Science* **287:** 1049–1053.
3. Bohnacker, T., R. Marone, E. Collmann, *et al*. 2009. PI3Kgamma adaptor subunits define coupling to degranulation and cell motility by distinct PtdIns(3,4,5)P3 pools in mast cells. *Sci. Signal.* **2:** ra27.
4. Stephens, L.R., A. Eguinoa, H. Erdjument-Bromage, *et al*. 1997. The G beta gamma sensitivity of a PI3K is dependent upon a tightly associated adaptor, p101. *Cell* **89:** 105–114.
5. Kurig, B., A. Shymanets, T. Bohnacker, *et al*. 2009. Ras is an indispensable coregulator of the class IB phosphoinositide 3-kinase p87/p110gamma. *Proc. Natl. Acad. Sci. USA* **106:** 20312–20317.

6. Del Prete, A., W. Vermi, E. Dander, *et al*. 2004. Defective dendritic cell migration and activation of adaptive immunity in PI3Kgamma-deficient mice. *EMBO J*. **23:** 3505–3515.

7. Camps, M., T. Ruckle, H. Ji, *et al*. 2005. Blockade of PI3Kgamma suppresses joint inflammation and damage in mouse models of rheumatoid arthritis. *Nat. Med*. **11:** 936–943.

8. Barber, D.F., A. Bartolome, C. Hernandez, *et al*. 2006. Class IB-phosphatidylinositol 3-kinase (PI3K) deficiency ameliorates IA-PI3K-induced systemic lupus but not T cell invasion. *J. Immunol*. **176:** 589–593.

9. Laffargue, M., R. Calvez, P. Finan, *et al*. 2002. Phosphoinositide 3-kinase gamma is an essential amplifier of mast cell function. *Immunity* **16:** 441–451.

10. Hirsch, E., O. Bosco, P. Tropel, *et al*. 2001. Resistance to thromboembolism in PI3Kgamma-deficient mice. *FASEB J*. **15:** 2019–2021.

11. Patrucco, E., A. Notte, L. Barberis, *et al*. 2004. PI3Kgamma modulates the cardiac response to chronic pressure overload by distinct kinase-dependent and -independent effects. *Cell* **118:** 375–387.

12. Chang, J.D., G.K. Sukhova, P. Libby, *et al*. 2007. Deletion of the phosphoinositide 3-kinase p110gamma gene attenuates murine atherosclerosis. *Proc. Natl. Acad. Sci. USA* **104:** 8077–8082.

13. Fougerat, A., S. Gayral, P. Gourdy, *et al*. 2008. Genetic and pharmacological targeting of phosphoinositide 3-kinase-gamma reduces atherosclerosis and favors plaque stability by modulating inflammatory processes. *Circulation* **117:** 1310–1317.

14. Fougerat, A., N.F. Smirnova, S. Gayral, *et al*. 2012. Key role of PI3Kγ in monocyte chemotactic protein-1-mediated amplification of PDGF-induced aortic smooth muscle cell migration. *Br. J. Pharmacol*. **166:** 1643–1653.

15. Vecchione, C., E. Patrucco, G. Marino, *et al*. 2005. Protection from angiotensin II-mediated vasculotoxic and hypertensive response in mice lacking PI3Kgamma. *J. Exp. Med*. **201:** 1217–1228.

16. Hotamisligil, G.S. 2006. Inflammation and metabolic disorders. *Nature* **444:** 860–867.

17. Solinas, G. & M. Karin. 2010. JNK1 and IKKbeta: molecular links between obesity and metabolic dysfunction. *FASEB J*. **24:** 2596–2611.

18. Solinas, G. 2012. Molecular pathways linking metabolic inflammation and thermogenesis. *Obes. Rev*. **13**, (Suppl 2): 69–82.

19. Kobayashi, N., K. Ueki, Y. Okazaki, *et al*. 2011. Blockade of class IB phosphoinositide-3 kinase ameliorates obesity-induced inflammation and insulin resistance. *Proc. Natl. Acad. Sci. USA* **108:** 5753–5758.

20. Becattini, B., R. Marone, F. Zani, *et al*. 2011. PI3K{gamma} within a nonhematopoietic cell type negatively regulates diet-induced thermogenesis and promotes obesity and insulin resistance. *Proc. Natl. Acad. Sci. USA* **108:** E854–E863.

Ann. N.Y. Acad. Sci. ISSN 0077-8923

ANNALS OF THE NEW YORK ACADEMY OF SCIENCES

Issue: *Inositol Phospholipid Signaling in Physiology and Disease*

Class III PI3K Vps34: essential roles in autophagy, endocytosis, and heart and liver function

Nadia Jaber and Wei-Xing Zong

Department of Molecular Genetics and Microbiology, Stony Brook University, Stony Brook, New York

Address for correspondence: Nadia Jaber, Stony Brook University–Microbiology, 225 Life Sciences Building, Stony Brook University, Stony Brook, NY 11794. nadiaj3@gmail.com

Mammalian phosphatidylinositol (PI) 3-kinases are a family of proteins that share the ability to phosphorylate phosphoinositides at the 3 position of the inositol ring. By doing so, these kinases produce phospholipid molecules that are involved in various cell signaling pathways, such as insulin signaling and endocytosis. The pathways regulated by PI3-kinases are crucial for maintaining cellular homeostasis and thus must be tightly regulated. Irregular PI3-kinase activity is observed in numerous human pathological conditions, such as diabetes, cancer, and inflammation. One family member, Vps34, is of particular interest because it is the only PI3-kinase identified in yeast and it has been evolutionarily conserved through mammals. Vps34 plays an essential role in the cellular process of autophagy, a process linked to human health and disease. Understanding the precise role of mammalian Vps34 will likely be integral to drug development for various diseases.

Keywords: Vps34; authophagy; endocytosis; PI3K; homeostasis

Vps34 in yeast

Vps34 was identified by a screen for genes involved in protein sorting to the vacuole in *Saccharomyces cerevisiae*. Subsequent studies revealed that the protein possessed PI3-kinase activity. Initial studies focused on the role of Vps34 in delivery of proteins such as hydrolases to the yeast vacuole, which is analogous to the mammalian lysosome. It was then discovered that yeast Vps34 forms at least two distinct macromolecular complexes, which separately regulate vacuolar protein sorting and macroautophagy.[1] Vps34 is believed to regulate these membrane trafficking events by producing PI(3)P, which recruits downstream effector proteins containing PI(3)P-binding domains.

Macroautophagy (also simply referred to as autophagy) is a catabolic cellular process involving the *de novo* formation of an organelle termed the *autophagosome*, which sequesters cytoplasmic material and degrades it via fusion with the vacuole/lysosome. Breakdown products, such as amino acids and lipids, are exported from the lysosome/vacuole into the cytoplasm for metabolic use. Autophagy is

constantly maintained at a basal level, whereby it functions as a quality control mechanism. However, it can also be enhanced in response to a number of stress signals, especially nutrient starvation. By degrading bulk cytoplasm or excess cellular material, autophagy provides the cell with required nutrients for survival. In addition, autophagy can be activated in response to hypoxia, DNA damage, and infection.[2] Because autophagy is essential for maintaining cellular homeostasis, its deregulation is observed in multiple diseases and may even contribute to pathogenesis.[3] Autophagy has a tumor suppressive role by maintaining cell health and homeostasis. However, evidence also shows that autophagy aids tumor cell survival following chemotherapy, and current efforts focus on combining traditional chemotherapeutics with autophagy inhibitors.

Vps34 also regulates protein delivery to the vacuole in yeast and may have a role in endocytosis in mammals. Similar to autophagy, endocytosis is a dynamic membrane trafficking pathway that helps maintain cellular homeostasis.[4] The early endosome serves to receive vesicular cargo from the plasma membrane and Golgi, and to sort the enclosed cargo

doi: 10.1111/nyas.12026

Ann. N.Y. Acad. Sci. 1280 (2013) 48–51 © 2013 New York Academy of Sciences.

for recycling or degradation. Once filled with cargo destined for degradation, the early endosome matures into a late endosome that then fuses with a lysosome to facilitate degradation. Vps34 is believed to produce PI(3)P on early endosomes and subsequently recruit downstream effectors that regulate vesicle docking and cargo sorting. Vps34 may also have a role in endosome maturation.

Discovery of a mammalian homolog

Efforts by Volinia *et al.* identified a human homolog of yeast Vps34p, with 58% sequence similarity.[5] Similar to yeast, mammalian Vps34 utilizes only phophatidylinositol (PI) as a substrate, thereby producing PI(3)P. Not long after, it was shown that feeding human colon cancer cells synthetic PI(3)P activates autophagy. It was also shown that PI3-kinase inhibitors wortmannin, LY294002, and 3-methyladenine (3-MA) block autophagy in mammalian cells.[6–8] Thus it seemed likely that Vps34 controlled mammalian autophagy via the production of PI(3)P. Likewise, many studies used these PI3-kinase inhibitors to describe a role for Vps34 at the early endosome.

However, mammals have evolved to encode multiple PI3-kinases with various activities and functions. The PI3-kinase inhibitors wortmannin, LY294002, and 3-MA are not selective for Vps34, and can target these other kinases. Therefore, it is difficult to draw definitive conclusions about the precise role of mammalian Vps34 in autophagy or endocytosis. Moreover, a recent study using a conditional deletion of *Vps34* in mouse sensory neurons found that the autophagic pathway was not disrupted in Vps34-deficient neurons.[9] These seemingly paradoxical observations prompted us to create our own conditional knockout model to further explore the precise role (or lack thereof) of mammalian Vps34 in autophagy.

The function of mammalian Vps34

Vps34 deletion in vivo

As whole-body knockout of Vps34 is embryonically lethal, we created a conditional knockout mouse using the Cre-lox system; Cre recombinase removes exon 4 of the Vps34 gene causing a frameshift and deletion of the majority of the protein. These animals were crossed with albumin-Cre or muscle creatine kinase (Mck)-Cre mice to delete Vps34 in the liver and heart, respectively (these organs

were chosen because autophagy has an integral role in their function). Mice with Vps34-deficient livers developed hepatomegaly and hepatosteatosis, as well as impaired protein turnover (Fig. 1A); similarly, mice with Vps34-deficient hearts developed cardiomegaly and displayed decreased heart contractility (Fig. 1B). Both conditional knockout mice begin dying around eight weeks of age.[10] Further characterization has shown that Vps34-null liver- and heart-specific mice resemble mice deficient in core autophagy proteins Atg5 or Atg7.

Effect on autophagy in vitro

To better understand the precise role of Vps34 in autophagy, we produced mouse embryonic fibroblasts (MEFs) from Vps34$^{f/f}$ mice after addition of adenovirus-containing Cre, which deletes endogenous Vps34. Vps34-null MEFs are devoid of GFP-FYVE puncta, an indicator of PI(3)P; we also observed a complete lack of autophagosomes in both untreated and serum-starved Vps34-null cells by electron microscopy (EM) (Fig. 1C); and the autophagic substrate p62 accumulates in Vps34-null cells. Compared with normal punctate structures in wild-type cells, the autophagosome marker LC3 surprisingly forms large aggregates in null cells. By immuno-gold EM, we determined that these LC3 structures are indeed cytoplasmic protein aggregates that are not associated with any membrane structures. These data suggest that autophagosome formation is completely blocked in the absence of Vps34.

Effect on endocytosis in vitro

We briefly explored the effect of Vps34 deletion on the endocytic pathway in our *in vitro* model. We found that endocytic recycling of transferrin, a function of the early endosome, is unaltered in Vps34-null cells. In comparison, endocytic degradation of the epidermal growth factor receptor (EGFR)—a function of the late endosome/lysosome—is severely abrogated. These data challenge the view that Vps34 is required to recruit sorting machinery to the early endosome.

Effect on mTOR signaling in vitro

Although it is known that mTOR is a potent inhibitor of autophagy, evidence also shows that Vps34 is required for the activation of mTOR in response to nutrient stimulation.[8] To evaluate the capacity for mTOR signaling in Vps34-null MEFs, we

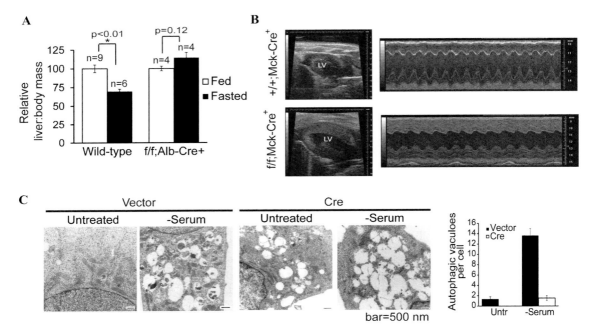

Figure 1. (A) Wild-type and Vps34$^{f/f}$-Alb-Cre$^+$ mice were fed normally or fasted for 24 hours. Liver to body mass ratio was determined and expressed as normalized to the respective fed animals. (B) Echocardiograms of wild-type and Vps34$^{f/f}$-Mck-Cre$^+$ animals *in vivo* demonstrates an increase in left ventricular wall thickness and mass, and decreased cardiac contractility. (C) Representative EM images of untreated or serum-starved Vps34$^{f/f}$ MEFs infected with empty vector or Cre. Serum starvation induces autophagosome formation in control but not Vps34-null cells. Large, single-membraned empty vacuoles are observed in both untreated and starved Vps34-null cells, likely representing enlarged late endosome/lysosomes. The data in this figure are from Ref. 10.

have begun studies using phosphorylation of downstream target S6 as a readout to determine whether amino acid–induced mTOR signaling is severely reduced in Vps34-null MEFs ; our initial results indicate that it is.

Summary

A member of the PI3-kinase family, Vps34 was originally identified in yeast by a screen for genes involved in protein sorting to the vacuole. It was later discovered that Vps34 also controls autophagy in yeast. The mammalian homolog was identified, and, based on the use of PI3-kinase inhibitors, a suggestion was made that the mammalian homolog also controls endocytic trafficking and autophagy in mammals. However, the PI3-kinase inhibitors currently available are not specific for Vps34, and it is therefore difficult to determine the precise contributions of Vps34 in these pathways. After a recent report that specific deletion of *Vps34* in mouse sensory neurons does not affect autophagy, we were prompted

to further investigate the role of mammalian Vps34 in autophagy.

We have produced liver- and heart-specific Vps34 conditional knockout mice. Characterization of the mice thus far shows that Vps34 is essential for normal heart and liver function; additional work *in vitro* shows that autophagosome formation is completely blocked in the absence of Vps34. We have also observed a surprising preservation of early endosome function, and a severe defect in late endosome function in the absence of Vps34. Finally, our work in progress indicates that Vps34 is necessary for amino acid–induced mTOR signaling. From these data we conclude that Vps34 is an important PI3-kinase responsible for autophagy induction in mammals. However, the precise role of Vps34 in the endocytic pathway requires further elucidation, as there are multiple differences between the effects of wortmannin and stable Vps34 deletion.

Vps34 inhibitors are currently under investigation because defects in autophagy have been implicated in the pathogenesis of multiple diseases. However, our work indicates that caution should be taken in these efforts, as systemic and/or complete inhibition of Vps34 will likely lead to severe cellular and organ damage. Rather than global inhibition of Vps3, then, small molecules that block partial functions of Vps34 (i.e., autophagy but not endocytosis) may be a better solution.

Conflicts of interest

The authors declare no conflicts of interest.

References

1. Kihara, A., T. Noda, N. Ishihara & Y. Ohsumi. 2001. Two distinct Vps34 phosphatidylinositol 3-kinase complexes function in autophagy and carboxypeptidase Y sorting in Saccharomyces cerevisiae. *J. Cell Biol.* **152:** 519–530.
2. Rabinowitz, J.D. & E. White. 2010. Autophagy and metabolism. *Science* **330:** 1344–1348.
3. Mizushima, N., B. Levine, A.M. Cuervo & D.J. Klionsky. 2008. Autophagy fights disease through cellular self-digestion. *Nature* **451:** 1069–1075.
4. Huotari, J. & A. Helenius. 2011. Endosome maturation. *EMBO J.* **30:** 3481–3500.
5. Volinia, S. *et al.* 1995. A human phosphatidylinositol 3-kinase complex related to the yeast Vps34p-Vps15p protein sorting system. *EMBO J.* **14:** 3339–3348.
6. Blommaart, E.F., U. Krause, J.P. Schellens, H. Vreeling-Sindelarova & A.J. Meijer. 1997. The phosphatidylinositol 3-kinase inhibitors wortmannin and LY294002 inhibit autophagy in isolated rat hepatocytes. *Europ. J. Biochem.* **243:** 240–246.
7. Petiot, A., E. Ogier-Denis, E.F. Blommaart, A.J. Meijer & P. Codogno. 2000. Distinct classes of phosphatidylinositol 3'-kinases are involved in signaling pathways that control macroautophagy in HT-29 cells. *J. Biol.l Chem.* **275:** 992–998.
8. Backer, J.M. 2008. The regulation and function of Class III PI3Ks: novel roles for Vps34. *Biochem. J.* **410:** 1–17.
9. Zhou, X. *et al.* 2010. Deletion of PIK3C3/Vps34 in sensory neurons causes rapid neurodegeneration by disrupting the endosomal but not the autophagic pathway. *Proc. Natl. Acad. Sci. USA* **107:** 9424–9429.
10. Jaber, N. *et al.* 2012. Class III PI3K Vps34 plays an essential role in autophagy and in heart and liver function. *Proc. Natl. Acad. Sci. USA* **109:** 2003–2008.

Ann. N.Y. Acad. Sci. ISSN 0077-8923

ANNALS OF THE NEW YORK ACADEMY OF SCIENCES
Issue: *Inositol Phospholipid Signaling in Physiology and Disease*

Inpp4b is a novel negative modulator of osteoclast differentiation and a prognostic locus for human osteoporosis

Jean Vacher

Clinical Research Institute of Montreal, Départment de Médecine, Université de Montréal, Department of Medicine, McGill University, Montréal, Québec, Canada

Address for correspondence: Jean Vacher, Clinical Research Institute of Montreal, 110 West Pins Avenue, Montreal H2W1R7 Quebec, Canada. vacherj@ircm.qc.ca

Inositol polyphosphate 4-phosphatase type II (Inpp4b) is a novel negative modulator of osteoclast differentiation and a prognostic locus for human osteoporosis. This short overview summarizes some of the cellular, molecular, and crosstalk signaling mechanisms that control osteoclast and osteoblast differentiation and activation.

Keywords: Inpp4b; osteoclast differentiation; bone mineral density; osteoporosis

Introduction

Bone homeostasis is controlled by two cellular processes: bone formation under the control of the osteoblast and bone resorption by the osteoclast.[1] Upon imbalance of one of these cellular functions, major bone defects, such as osteopetrosis, and osteoporosis, can occur.[2] Hence, deciphering mechanisms that control these two cellular processes is crucial for exploring new approaches toward treatment of these clinically important diseases. Our goals are to understand the cellular, molecular, and crosstalk signaling mechanisms that control osteoclast and osteoblast differentiation and activation. Recently, our studies focused on the signaling mechanisms in osteoclast differentiation and discovered that inositol polyphosphate 4-phosphatase type II (Inpp4b) plays a key molecular and physiologic role in bone homeostasis.

The PI3-kinase signaling pathway in bone biology

The phosphatidylinositol 3-kinase (PI3K) pathway has been involved in numerous regulatory mechanisms, including cell growth, differentiation, and survival.[3] Some previous studies suggested a major role for the PI3K pathway[4,5] in bone physiology, in particular two inositol phosphatases that antagonize PI3K action. *Pten*, the phosphatase and tensin homolog deleted on chromosome 10, is a tumor suppressor gene and encodes an inositol 3-phosphatase essential for osteoblast biology. Loss of function of Pten leads to a high bone mass phenotype in the mouse with increased osteoblast survival.[6] In contrast, ablation of the SHIP gene (*Inpp5d*), the SH2 domain-containing inositol 5-phosphatase, in the mouse resulted in a low bone mass phenotype or osteoporosis due to stimulation of bone resorption by the osteoclasts.[7] The inverse physiologic response of Pten and SHIP deficiency is attributable to the two opposite bone cell functions. These experiments also emphasize the highly sensitive regulatory mechanism and the importance of crosstalk between osteoblast and osteoclast. A third inositol phosphatase, Inpp4b, previously identified, was proposed based on genomic analysis to be a tumor suppressor in human cancers including breast, prostate, and ovaries.[8] Recently, we demonstrated that Inpp4b plays a major role in bone physiology by antagonizing the PI3K pathway in osteoclasts.[9]

Osteoclast differentiation is modulated by Inpp4b *ex vivo* and *in vivo*

We have isolated and characterized the mouse Inpp4b gene and structure.[10] Inpp4b is preferentially expressed in hematopoietic tissues and, most

doi: 10.1111/nyas.12014

 Ann. N.Y. Acad. Sci. 1280 (2013) 52–54 © 2013 New York Academy of Sciences.

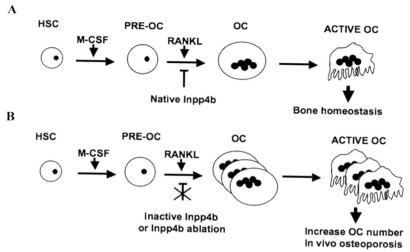

Figure 1. Role of Inpp4b in osteoclast differentiation. (A) In physiologic conditions, native Inpp4b is a negative modulator of osteoclast differentiation to control bone homeostasis with the M-CSF and RANKL cytokines. (B) Ectopic expression of the catalytic inactive form of Inpp4b or ablation of Inpp4b in osteoclasts stimulates osteoclast differentiation and leads to an increase in osteoclast number, more resorption, and osteoporosis in mice. HSC: hematopoietic stem cell; PRE-OC: preosteoclast; OC: osteoclast; M-CSF: macrophage colony-stimulating factor; RANKL: receptor activator of nuclear factor-κB (NF-κB) ligand.

abundantly, in the osteoclast lineage, whereas expression is not detectable in the osteoblast. To define the role of Inpp4b in bone physiology and specifically in the osteoclast, various *ex vivo* and *in vivo* complementary approaches were undertaken.

First, we investigated the role of Inpp4b by using the monocytic RAW cell line that can differentiate into multinucleated mature osteoclasts. Stable expression of the native Inpp4b in RAW cells led to poor differentiation and thereby produced fewer mature osteoclasts compared to controls. These results suggested that Inpp4b is a negative modulator of osteoclastogenesis. Importantly, this function is dependent on the catalytic activity of Inpp4b since expression of the inactive form of Inpp4b in RAW cells resulted in a major increase of the mature osteoclast population. Interestingly, these mature osteoclasts were significantly larger in size. This finding suggested a potential additional role of Inpp4b in osteoclast cytoskeletal remodeling, essential for proper bone resorption.

Second, we conducted *in vivo* studies on the role of Inpp4b in osteoclast differentiation, by ablation of the Inpp4b gene in the mouse. Mice deficient for Inpp4b systemically do not show major developmental defects and have normal life span. Upon careful phenotypic characterization, however, these mice demonstrated a significant reduction of bone

mass. This osteoporotic phenotype was associated with an increase in the mature osteoclast population *in vivo*. These experiments supported and confirmed Inpp4b's role as a negative modulator of osteoclastogenesis. In parallel, loss of Inpp4b function in these mice induced an increase of the osteoblast population as a secondary response to osteoclast lineage stimulation. Significantly, this finding indicated that the osteoclast/osteoblast crosstalk mechanism is unaffected in the absence of Inpp4b. Most importantly, these experiments demonstrate that Inpp4b is a negative modulator of osteoclast differentiation both *ex vivo* and *in vivo*.[9]

Inpp4b modulates Nfatc1 signaling in the osteoclast

To define the pathway(s) potentially modulated by Inpp4b, we analyzed all known intracellular signaling pathways involved in osteoclast differentiation. These include the MAP kinases (p38, JNK, Erk1/2), Akt, NF-κB, Nfatc1, and phospholipase C gamma (PLCγ). In RAW cell–derived osteoclasts, expression of the inactive form of Inpp4b induced a slight increase in activation of NF-κB and Akt proteins and no detectable stimulation for PLCγ. In contrast, the Nfatc1 pathway showed a strong response with a major relocalization of the Nfatc1 transcriptional factor to the nucleus.

Similar signaling responses were also evaluated in mature osteoclasts derived from Inpp4b$^{-/-}$ mice, which implicated Inpp4b in regulation of the Nfatc1 pathway.

A key role for Inpp4b in modulation of Nfatc1 was substantiated in RAW cell–derived osteoclasts that showed enhanced intracellular osteoclast calcium oscillations caused by increased calcium release from intracellular stores. Upon high cytosolic calcium concentration, the phosphatase calcineurin is activated and can dephosphorylate the cytosolic inactive form of Nfatc1. This dephosphorylation induces, in turn, relocalization of Nfatc1 to the nucleus that then stimulates transcription of osteoclast-specific Nfatc1 target genes, including cathepsin K, $\alpha_V\beta_3$ integrins, TRAP (Acp5), and Nfatc1 itself.[9] Hence, modulation of the Nfatc1 pathway in osteoclasts either deficient for Inpp4b or expressing the inactive form of Inpp4b protein results in stimulation of osteoclast differentiation and, ultimately, an increase in the active osteoclast population. *In vivo*, this mechanism leads to significant upregulation of bone resorption and osteoporosis in Inpp4b-null mice (Fig. 1).

INPP4B as a bone mineral density variability marker in humans

Based on our results in the mouse and given that we localized Inpp4b in the mouse genome to a syntenic human chromosomal region, including bone mineral density quantitative trait loci (QTL),[10] we investigated *INPP4B* as a candidate gene. Two cohorts of healthy premenopausal women from Toronto and Quebec City of Caucasian origin were genotyped by high throughput sequencing for 83 single nucleotide polymorphisms associated with the *INPP4B* locus. Importantly, three INPP4B variants in our analysis showed a significant association with bone mineral density variability in these populations.[9] These results revealed that INPP4B could be a potential contributor to osteoporosis in humans. In fact, values for INPP4B bone mineral density variability were comparable to those of LRP5, the well-recognized genetic trait for bone mineral density.

In summary, work from my laboratory has characterized the lipid phosphatase Inpp4b, a member of the PI3K pathway, as a novel modulator of bone mass in mice and humans. These findings uncovered key insights into our understanding of the molecular regulatory mechanism of osteoclast physiology. Importantly, these studies may also affect the human population by providing an early genetic prognosis for the healthy individuals at risk to develop osteoporosis.

Acknowledgments

This work was supported by the Canadian Institutes of Health Research (CIHR) Grant MOP93639, and by the Natural Sciences and Engineering Research Council of Canada (NSERC).

Conflicts of interest

The author declares no conflicts of interest.

References

1. Karsenty, G. 2003. The complexities of skeletal biology. *Nature* **423:** 316–318.
2. Zaidi, M. 2007. Skeletal remodeling in health and disease. *Nat. Med.* **13:** 791–801.
3. Cantley, L.C. 2002. The phosphoinositide 3-kinase pathway. *Science* **296:** 1555–1557.
4. Golden, L.H. & K.L. Insogna. 2004. The expanding role of PI3-kinase in bone. *Bone* **34:** 3–12.
5. Guntur, A.R. & C.J. Rosen. 2011. The skeleton: a multifunctional complex organ. New insights into osteoblasts and their role in bone formation: the central role of PI3Kinase. *J. Endocrinol.* **211:** 123–130.
6. Liu, X., K.J., Bruxvoort, C.R. Zylstra, *et al.* 2007. Lifelong accumulation of bone in mice lacking Pten in osteoblasts. *Proc. Natl. Acad. Sci. USA* **104:** 2259–2264.
7. Takeshita, S., N. Namba, J.J. Zhao, *et al.* 2002. SHIP-deficient mice are severely osteoporotic due to increased numbers of hyper-resorptive osteoclasts. *Nat. Med.* **8:** 943–949.
8. Agoulnik, I.U., M.C. Hodgson, W.A. Bowden, *et al.* 2011. INPP4B: the new kid on the PI3K block. *Oncotarget* **2:** 321–328.
9. Ferron, M., M. Boudiffa, M. Arsenault, *et al.* 2011. Inositol polyphosphate 4-phosphatase B as a regulator of bone mass in mice and humans. *Cell Metab.* **14:** 466–477.
10. Ferron, M. & J. Vacher. 2006. Characterization of the murine Inpp4b gene and identification of a novel isoform. *Gene* **376:** 152–161.